# CONTEMPORARY CASE STUDIES

# Rivers & Coasts

## Bob Hordern

Series Editor: Sue Warn

Philip Allan Updates
Market Place
Deddington
Oxfordshire
OX15 0SE

Tel: 01869 338652
Fax: 01869 337590
e-mail: sales@philipallan.co.uk
www.philipallan.co.uk

Front cover photograph reproduced by permission of Kitchenham Ltd.

Printed by Raithby, Lawrence & Co. Ltd, Leicester

### Environmental information
The paper on which this title is printed is sourced from managed, sustainable forests.

# Contents

## Introduction

## Part 1: Rivers and systems .........................................................1

## Part 2: River landforms and processes ................................10

## Part 3: Changing river channels ...........................................21

## Part 4: Rivers and people ......................................................30

# Introduction

Rivers and coasts are dynamic environments within which the processes of erosion, transportation and deposition create and change landforms. Key elements in both of these environments are the movement of water and sediments.

Drainage basins form the main element of river landscapes and, viewed from a hydrological perspective, this means the input, throughput and output of water. The fluvial system also includes the movement of sediment and the creation of landforms. The interaction of variables such as river discharge, sediment and rock type lead to natural changes in river channels.

The main focus in coastal environments is the littoral zone. In this part-marine, part-terrestrial environment, wave energy and shoreline geology are key variables. Many technical terms apply to both coasts and rivers, although differential erosion and longshore drift are important along coasts. Ecosystems such as sand dunes emerge as distinctive features of many coastlines.

Links between rivers and coasts are important and most noticeably seen in large rivers, where deltas, mudflats, ecosystems and beaches are evidence of these interactions. River processes are themselves partly controlled by sea level changes. It is often only when human activities interfere with natural processes that the importance of these links becomes apparent.

River floodplains and coastlines are among the most densely populated areas of the world. It is therefore not surprising that both have become battlegrounds where natural processes and human technology often face each other head on. Conflicts over land use occur against a background of a rise in global sea level, more frequent storms and unexpected flood events.

River and coastal management strategies range from traditional hard engineering to softer approaches that work with nature. River flood protection uses afforestation and land use planning, as well as dams and channelisation, while beach nourishment is used to support sea wall and groyne construction along coastlines.

Sustainable management strategies are being increasingly implemented to tackle erosion, flooding, water management and pollution issues. River restoration and managed coastal retreat are examples of recent policies being developed.

## About this book

This book covers a range of topics found in most A-level courses and is divided into 11 parts: five on rivers, five on coasts and one on examination advice. Key concepts are explained and supported by case studies that illustrate the issues involved and develop understanding.

Questions are provided throughout, focusing on data–response, which check your understanding and allow you to practise important skills. Many questions are in a format designed to help with your revision or examination techniques.

The last section offers more detailed advice on how best to use case studies and provides a bank of examination essay titles to analyse, plan and use.

# Key terms

These have been divided into four groups:

### Systems
**Cliff cycles** illustrate the way in which resistant coastlines develop a series of characteristic landforms

**Hydrographs** show how river discharge changes over short periods of time.

**Hydrology** is the study of how water moves through the landscape. Inputs of precipitation are variously affected by interception, infiltration, runoff and evapotranspiration.

**Littoral cells** are convenient areal units within which shorelines can be studied and managed.

**River regimes** show how rivers respond to seasonal changes in weather, or the effects of tributaries.

**Succession** and **zonation** are ways in which coastal ecosystems, such as sand dunes, show the effects of change.

**Water budget graphs** show how seasonal changes in soil moisture relate to climate and vegetation changes.

### Landforms
**Beaches** are not simple coastal features and can vary in terms of the calibre of their deposits, their various shapes and the way in which they are constructed. They include swash and drift-aligned forms and have complex surface and linear features such as spits and bars.

**River channels** may be studied in terms of their cross-sections or their sinuosity, (e.g. meandering or braided).

**River long profiles** are downstream transects of rivers, showing the effects of erosion and deposition, leading to characteristic landforms. Local geology may interrupt this profile, creating waterfalls or lakes.

**Structure** and **lithology** are two characteristics of rocks that have a profound effect upon cliff coastlines. They create a series of distinct landforms.

**Valley cross profiles** illustrate the effects of vertical and lateral erosion. They include both the valley sides and river channels.

## Processes

**Deposition** is linked to the power or velocity of water, whether in stream channels or waves. Zones of sedimentation, or sinks, include floodplains, deltas, mudflats and beaches.

**Differential erosion** occurs when differing rock types (often next to each other) produce varied rates of erosion. This is most noticeable along coastlines.

**Erosion** includes a series of processes including abrasion (corrasion), hydraulic action, attrition and solution which operate in river and marine environments. The agents involved may be rivers or waves.

**Eustatic** (worldwide) and **isostatic** changes in the relative levels of land and sea. These can have climatic and tectonic origins and give rise to characteristic landforms, such as fjords or raised beaches.

**Hjulström curves** are a way of relating sediment size to the processes of erosion, transportation and deposition.

**Marine processes** are influenced by wave characteristics, tides, storm surges and changes in sea level.

**Mass movement** removes weathered material down slope under gravity. Characteristic types include falls, creeps, slides and slumps. These reflect the speed of movement and the role of water.

**River channel variables** are important factors when trying to understand how rivers work. They include velocity, discharge, channel shape and load.

**Transportation** moves material from one location to another, especially by water. This may occur in river or marine environments and involves processes such as traction, saltation and suspension. Water may also move in distinct ways, for instance, helically in rivers or in wave oscillation.

**Weathering** breaks rocks down *in situ*, mechanically, chemically and biologically.

## Management

**Coastal** and **river management** involves both policy and method/technique. Decisions are based on cost-benefit or environmental impact analyses.

**Flood hazards** occur across floodplains and along coasts when excessive amounts of water threaten people or property. They may have natural and human causes.

**Hard engineering** involves building defences, for example, dams, levees or seawalls. These are designed to resist or overcome the impacts of flooding or erosion.

**River restoration** and **managed retreat** are relatively recent strategies that attempt to deliver sustainable solutions in lowland environments.

**Soft engineering** tries to work with nature, copying or even using natural processes. Allowing some land to flood in order to protect other locations is seen in floodplain and wetland management plans.

**Sustainable management** attempts to meet current demands without compromising the needs of future generations.

# Rivers and systems

## Hydrological pathways

Before exploring landforms, processes and management strategies, it is important to have a wider view of rivers and systems. The way in which water and sediment move through the landscape is controlled by the **hydrological system** and the nature of the **drainage basins** involved. River basins form perhaps the most common element in the world's landscapes, despite being in places masked by vegetation or human development. These open systems are convenient units for study, having clear boundaries (watershed and coast) and a range of quantifiable **inputs**, **stores**, **flows** and **outputs** (see Figure 1).

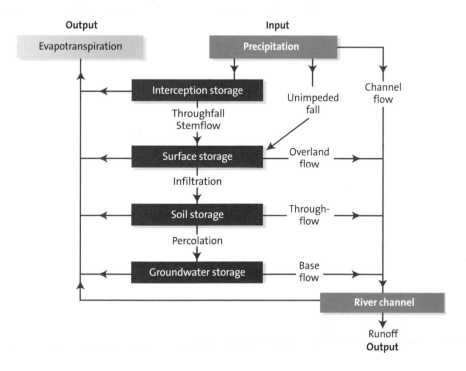

*Figure 1*
*How water reaches the river*

Closed hydrological systems involve the input of water into the catchment as precipitation or snow melt, its subsequent passage into river channels and its eventual return to the atmosphere either locally or via the oceans.

It is important to understand how and why water travels through the landscape. The proportion of water that becomes overland flow is the variable that most influences **runoff**, **erosion**, **river discharge** and **load**.

## Inputs

The impacts of heavy precipitation on river catchments — particularly with regard to downpours or melting snow — have been widely documented, although persistent rainfall can produce problems in some catchments. Antecedent conditions — what has happened previously — are also important when examining how rivers might respond to such events.

## Outputs

The main outputs from the system are **evapotranspiration** and runoff. **Evaporation** takes place from land and water surfaces, while transpiration releases water from leaf pores.

- Potential evapotranspiration is calculated assuming there is always sufficient water available. Where it exceeds precipitation, there will be a water deficit; higher precipitation leads to a water surplus.
- Runoff along river channels provides the means by which remaining water leaves the system. The basic equation used by hydrologists when examining river systems is:

$$\text{runoff} = \text{precipitation} - \text{evapotranspiration} \pm \text{storage}$$

## Stores

Water is found above and below ground in a variety of stores. These receive, hold and release water in response to the system. However, it is important to remember that 94% of global water is stored in the oceans, and much of the rest is ice.

- Interception by vegetation and buildings can prevent as much as 40% of precipitation reaching the ground in summer in the UK.
- Surface storage occurs in natural features such as lakes, reservoirs and wetlands, but also in urban areas because of impermeable surfaces.
- River channels also hold a lot of water temporarily, especially when in flood.
- Soil moisture levels depend upon soil texture and structure. Soils may be free-draining (gravel); others may absorb water (crumb structure) and in the case of fine (clay) soils become easily saturated.
- Groundwater accounts for 25% of the Earth's fresh water and is the major supplier of river flow. Below the water table, water is trapped in the pores and joints within rocks (aquifers). Some deeper stores may be regarded as fossil water as they are no longer part of the cycle.

## Flows

This is when water is transferred between the various stores in the system.

- Natural flows within vegetation cover include **stemflow** and drip. These are relatively slow processes compared with artificial pipes and gutters on buildings.

- **Infiltration** and **percolation** rates depend on permeability — this is one reason why geology can be a significant local factor in hydrology.
- **Overland flow** occurs when the ground is impermeable or if it has become saturated. Heavy rainstorms and snow melt often trigger such conditions.
- **Throughflow** is less important. It moves laterally, mainly along soil horizons.
- **Baseflow** in rivers is supplied by groundwater flow rather than rainfall.

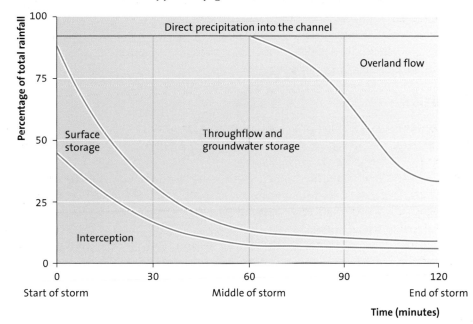

*Figure 2*
*How hydrological pathways change during a storm*

**1 Using case studies**

## Question

Using Figure 2, explain what happens to the various processes shown as the storm continues.

## Guidance

Be careful how you read this type of graph. Describe the changes by giving some figures, and then say why each change happens by completing the following:

*At the start, interception and surface storage each account for about 45% of rainfall, while most of the remaining 10% falls straight into the stream. Rainfall is collecting on the leaves of vegetation or remaining on the ground surface as puddles.*

*However, after 60 minutes…*

*By the end of the storm…*

Relationships between hydrological variables can be complex. Climate, relief and vegetation are the more obvious natural factors at work, but there are also changes that result from interference by people. Clearing vegetation, building towns and developing water management strategies alter the balance of the system. Figure 3 shows what happens to the water supply on the Mediterranean island of Cyprus.

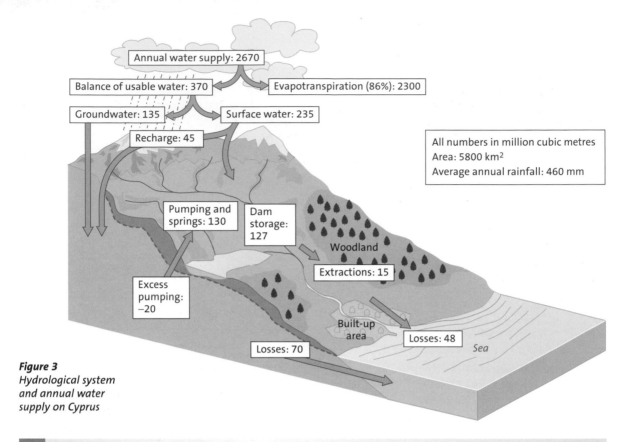

**Figure 3**
*Hydrological system and annual water supply on Cyprus*

Labels within the figure:

Annual water supply: 2670
Balance of usable water: 370
Evapotranspiration (86%): 2300
Groundwater: 135
Surface water: 235
Recharge: 45
Pumping and springs: 130
Dam storage: 127
Woodland
Extractions: 15
Excess pumping: −20
Built-up area
Losses: 48
Sea
Losses: 70

All numbers in million cubic metres
Area: 5800 km²
Average annual rainfall: 460 mm

## Question

Study Figure 3.
(a) Describe what happens to the island's 2670 million cubic metres of annual water supply.
(b) Identify two ways in which Cyprus tries to manage its water supplies.
(c) Why are water management problems greater in the summer?

## Guidance

These questions relate to how water moves throughout the landscape, naturally and artificially. Trace the paths and use of water above and below ground. Water management involves both supply and demand.

# River flow

The arrival of water in rivers causes changes in depth, speed and flow, which can be investigated using fieldwork or monitored remotely with automatic equipment. These data improve our understanding of river systems and give an insight into how best to manage issues such as flooding and water use. Discharge (calculated by multiplying the area of channel cross-section by downstream velocity) varies spatially through the basin and over time. Channel runoff can be examined in both the short and long term.

# Storm hydrographs

Storm **hydrographs** are graphs that show changes in a river's discharge during and after a storm. Their shape is determined by the different routes water takes through the landscape. Hydrographs show the changing discharge measured in cumecs (cubic metres of water per second).

When analysing these graphs, it is useful to use the terminology in Figure 4. You may be asked to compare graphs of the same river on different occasions or different rivers responding to the same storm. Key points to mention when describing hydrographs are:

- the steepness of the rising and falling limbs
- the value of the peak discharge
- the lag time (the period of time between peak rainfall and peak discharge)
- the time taken to return to normal

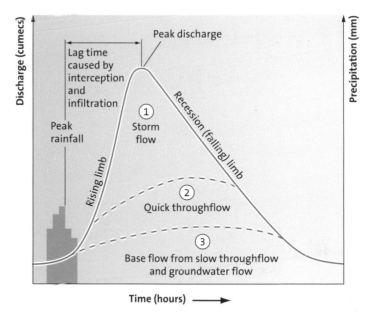

**Figure 4**
*A storm hydrograph*

## THE HYDROLOGY OF AUSTWICK AND CLAPHAM BECKS

*Case study 1*

Austwick Beck and Clapham Beck drain the southern slopes of Ingleborough in the Yorkshire Dales. They show how geology and land use can cause different responses to the same storm.

- Austwick Beck rises in Crummackdale where the rocks are grits, shales and flagstones. These rocks are impermeable and so give rise to many surface streams over a wide area. The **catchment area** is smaller than for the more elongated Clapham Beck.

**Figure 5** *Upstream along Austwick Beck*

**Figure 6** *Limestone moorland above Clapham Beck*

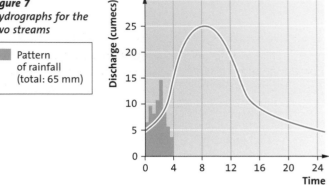

**Figure 7**
*Hydrographs for the two streams*

| Pattern of rainfall (total: 65 mm) |

**(A) Austwick Beck**

Discharge (cumecs)
25
20
15
10
5
0
0 4 8 12 16 20 24
Time
(hours from start of rainstorm)

**(B) Clapham Beck**

Discharge (cumecs)
25
20
15
10
5
0
0 4 8 12 16 20 24
Time
(hours from start of rainstorm)

Ingleborough
724 m ▲

Gaping Gill pothole

Clapham Cave

Clapham ●

B

157 m ■

A

● Austwick

N

Gritstone and sandstone

Carboniferous limestone

Shales

Planted woodland

Streams and artificial lake

Fault line

● Village

■ Spot height

▲ Peak

= Stream gauge

0   km   2

**Figure 8**
*Sketch map showing details of the two catchments*

■ Clapham Beck drains the slopes of Ingleborough as Fell Beck. The stream runs over moorland until it reaches the pervious Carboniferous limestone rock falling over 125 m into the pothole of Gaping Gill. Water travels underground through this resistant but pervious rock and, after some days, appears from Clapham (Ingleborough) Cave. From there it flows through a planted woodland and then into an artificial lake before reaching the gauging station at Clapham.

## Question

Study Figure 7. The hydrographs show how two neighbouring streams in North Yorkshire responded to the same rainstorm.

(a) Use hydrograph A (Austwick Beck) to work out the following values:
  (i) the peak discharge in cumecs
  (ii) the discharge in cumecs after 24 hours
  (iii) the lag time in hours
(b) Compare how the two streams responded to the rainstorm.

**Using case studies 3**

### Guidance

**(a)** Lag time is the delay between peak rainfall and peak discharge.

**(b)** Analysing another student's answer is one way to understand the ideas in a case study. Here is an example: *Austwick Beck (hydrograph A) has a flashy response to the storm, with steeper rising and falling limbs than hydrograph B. The peak of the graph is higher at 25 cumecs after a lag time of only 6 hours. Clapham Beck (hydrograph B) has a slower response with a peak of just 10 cumecs after nearly 17 hours. The limbs are much gentler and the stream is slower to return to normal.*

### Question

The sketch map in Figure 8 provides information about the drainage basins of the Austwick Beck and Clapham Beck. Using evidence from the map, explain how physical and human factors led to differences in the shapes of the hydrographs in Figure 7.

### Guidance

Here is another sample answer for you to complete:

*Physical factors include geology. The hydrograph for Austwick Beck (hydrograph A) is caused by impermeable rocks that encourage quick runoff and lots of small surface streams. The longer Clapham Beck has a slower response that is also related to the rock type: limestone is permeable, meaning more infiltration and percolation, and so lengthening the lag time. The underground stream slows down the response even further. Human factors at work are…*

**Using case studies 4** (margin)

# River regimes

A **river regime** describes the longer-term pattern of variations in discharge. These variations can be plotted on a graph to form a year-long hydrograph (Figure 9). This hydrograph may show marked seasonal peaks and low flows, greatly influenced by changes in precipitation, temperature, vegetation or geology. For example, big swings in discharge in tropical rivers relate to the wet and dry seasons, spring increases often suggest melting snow and permeable rocks reduce discharge most of the year.

A
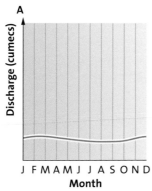

B
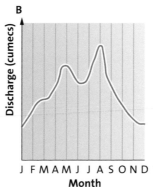

C

**Figure 9**
*Three river regimes*

## Question

(a) Define the term 'regime'.

(b) Study Figure 9. Match the following descriptions to rivers A, B or C.
- A large river in the Alps fed by snowfields and glaciers.
- A small river in southern England flowing on chalk rock.
- A large river flowing into the Mediterranean Sea.

(c) Explain your decisions.

## Guidance

These questions show the importance of climate in determining river flow, and how discharge changes seasonally.

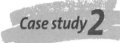

## THE VARYING REGIME OF THE RIVER RHÔNE

The River Rhône, at over 800 km in length, is one of the major rivers in Europe. The Rhône runs from Switzerland into France. It rises near the Rhône glacier at an altitude of 1753 m and flows west through Lake Geneva before entering France. At Lyons it is joined by its largest tributary, the Saône, before turning south between the mountains of the Massif Central and the Alps. Other alpine tributaries then enter this middle section of the Rhône, including the Isère and the Durance. Below Beaucaire, the Rhône divides into two branches to form the Camargue delta, before flowing into the Mediterranean Sea (Figure 10).

In common with many large rivers, the regime of the Rhône shows varying patterns in each section of the river, largely as a result of differing local climates. These variations can be seen in the graphs in Figure 10.

- The Swiss Rhône is an alpine stream. The regime shows low levels in winter, with increases in spring and summer caused by melting snow and glaciers. This pattern is partly smoothed out by Lake Geneva, which regulates the river's flow.
- Between Geneva and Lyons, the streams that join the Rhône are also alpine, increasing the marked spring peak.
- At Lyons, rain from Atlantic air masses falling in the Saône catchment creates an almost opposite pattern of flow in this tributary. The regime has high autumn and winter flows, with a clear drop in summer.
- After the confluence of the two rivers at Givors, the combined regime produces a large overall increase with a maximum flow at the end of winter.
- Further south, while the summer evaporation reduces discharge, meltwater from French alpine tributaries like the Isère restores the spring maximum.
- At Beaucaire, the mean discharge reaches 1683 cumecs, with high spring and autumn flows from Mediterranean rainfall. A figure as high as 13 000 cumecs has been recorded, causing floods in December 2003.

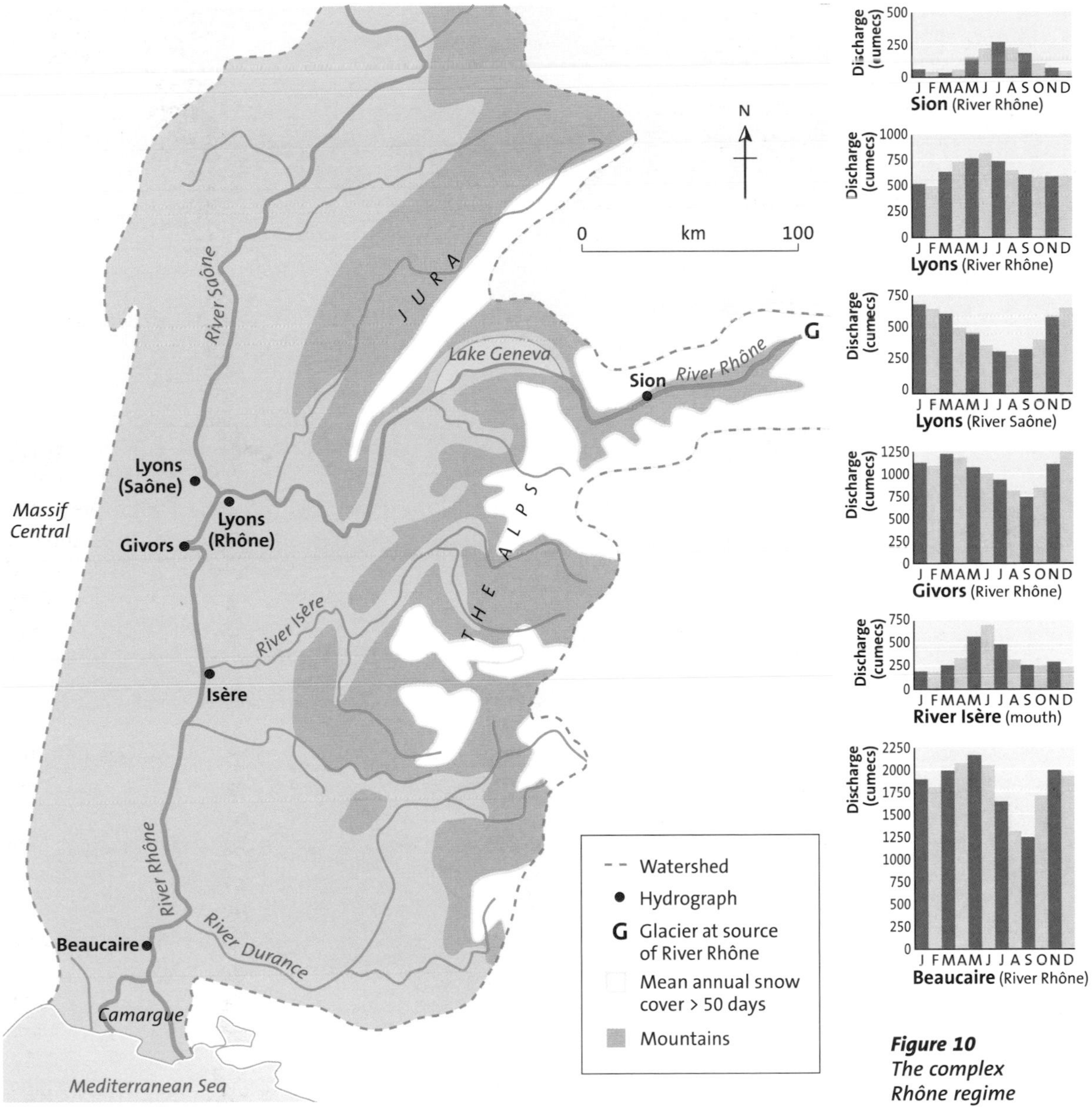


***Figure 10***
***The complex Rhône regime***

## 6 Using case studies

### Question

**(a) List the factors that cause changes to the regime of the River Rhône catchment.**

**(b) Suggest why the shapes of the two graphs at Lyons are different.**

**(c) Using Figure 10, annotate a sketched copy of the graph for Beaucaire to explain the annual pattern of discharge.**

### Guidance

Identify the peaks and troughs, annotating each to suggest why they occur. They result from what happens upstream and are influenced by seasonal factors and tributaries joining the river.

# Part 2
# River landforms and processes

## Recognising river landforms

River landforms change downstream as the processes of erosion, **transportation** and **deposition** interact. Figure 11 identifies three characteristic zones of landforms.

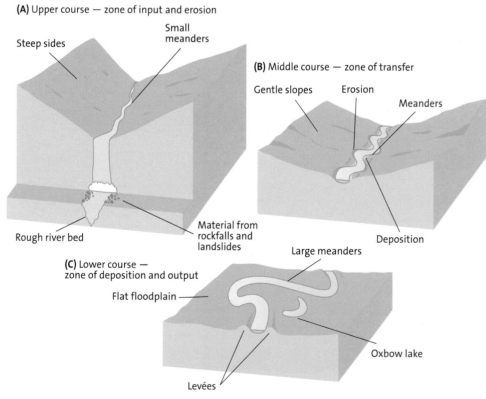

**(A)** Upper course — zone of input and erosion

Steep sides

Small meanders

Rough river bed

Material from rockfalls and landslides

**(B)** Middle course — zone of transfer

Gentle slopes

Erosion

Meanders

Deposition

**(C)** Lower course — zone of deposition and output

Large meanders

Flat floodplain

Oxbow lake

Levées

*Figure 11*
*River landforms —*
*a checklist from GCSE*

- The **upper course** of a river has inputs of water and material from mass movement, such as rockfalls and landslides. The river is only able to remove a small amount of its load, which is often angular and large. Geology may be important locally, helping form waterfalls and gorges. The main process is erosional cutting down into the river channel. Characteristic features include steep slopes, V-shaped valleys and interlocking spurs. River channels are narrow and the river bed is rough with a steep downstream gradient.

- The **middle course** of a river is a zone of transfer, where water and material brought from upstream are transported downstream. Increasing amounts of discharge (from tributary streams) and sediment are involved. The river's load becomes smaller in size and more rounded (the effect of **attrition**) and is carried in **suspension** or moved as **bedload**. Erosion occurs laterally on the outer banks of meanders and deposition on the inner. Meanders, floodplains and assymetrical channels are characteristic features. The river channel is smoother than upstream and the downstream gradient is much reduced.

- The **lower course** is a zone of mainly deposition and output. Within the channel, the sediment is sorted by the fastest current. In times of lower discharge, the coarser gravels collect as deposits (braiding). In times of flood, both water and silt (alluvial material) travel overbank onto the floodplain, building levées alongside the river. Eventually, deposition may block off old meanders to form oxbow lakes. This has the effect of straightening the river's course.

# Shaping valley cross profiles

Valley sides are shaped by a series of processes that provide material and move it downslope, ultimately to be transported by rivers. The main **sub-aerial processes** involved are **weathering** and **mass movement**.

## Weathering

This involves the breaking down of material *in situ*. There are three main types of weathering:

- **Mechanical (physical) weathering** is commonly caused by frost. Water enters joints or weaknesses and, on freezing, expands to shatter the rock. Collections of angular particles are often seen along mountain ridges or form scree on steep hillsides. Rocks with a massive structure are more easily shattered. In tropical deserts, intense heat may damage bare rock surfaces.

- **Chemical weathering** can take various forms, but the most obvious is found in limestone areas. Rainwater contains carbon dioxide in solution. This weak acid reacts with the calcium carbonate in limestone, making it soluble in water. Hydrolysis and oxidation are other processes of this sort.

- **Biological weathering** occurs when tree roots open up rock crevices (resulting in mechanical weathering) or decaying vegetation releases acids (resulting in chemical weathering).

## Mass movement

This is the subsequent movement downhill of the material provided by weathering. Rock type, climate, vegetation and the availability of water help to explain the scale and speed of these movements. Gravity is the essential force involved. Understanding how some of these processes work is more important than trying to classify all of the various types possible, so four typical examples are considered here:

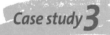

**Figure 12**
*Riverside soil creep and landslide*

- **Rockfalls** tend to occur from cliffs, producing rocky outcrops or scars with scree below. There is little water involved.
- **Slides** happen suddenly. Water soaks into the ground and the increased weight triggers movement along a now lubricated slip plane. Weak rocks, such as clay, have a tendency to rotate backwards when slipping and produce a series of concave steps. If the material becomes saturated, the landslide may lose its form and create a slump of material.
- **Flows** involve considerable quantities of water. **Mudflows** are a common hazard on steep tropical hillsides, volcanic slopes and old mining tips. They sometimes lead to tragic events.
- **Creeps** are slow, but continual, processes by which fine material is moved. Evidence on slopes includes leaning trees and walls, building subsidence and soil terracettes (Figure 12).

Surface water can also have an impact, although this is perhaps more properly classed as an erosion process. Rain may wash material downslope, especially when water cannot infiltrate. This is a common occurrence if there is intense rainfall, snow melts rapidly or the ground becomes saturated. When this happens, water runs downslope or collects in small rills. If there is little surface vegetation or the slopes are steep, these may grow to form gullies. At the foot of valley slopes, any loose material deposited is easily transferred into the river system.

*Case study 3*

## INVESTIGATING GORDALE

The valley of Gordale Beck (Figure 13) is close to Malham Cove in the Yorkshire Dales. From the foot of a waterfall, the stream emerges into a gorge with steep cliffs on either side.

**Figure 13**
*Sub-aerial processes shaping the valley sides of Gordale Beck near Malham*

*Contemporary Case Studies*

These limestone rocks, known as scars, show the effects of weathering, a process helped by high winter rainfall and cold night-time temperatures. Limestone also has distinct joints and bedding planes, which are easily exploited by all three types of weathering process.

Mass movement here depends on the size and shape of the rock pieces rather than the presence of water. The lower valley sides are made up of scree that builds up as rocks fall from above. Scree materials creep slowly downhill, often showing clear patterns in size and orientation.

Some erosion takes place in winter, when increased stream discharge may attack the foot of the valley sides.

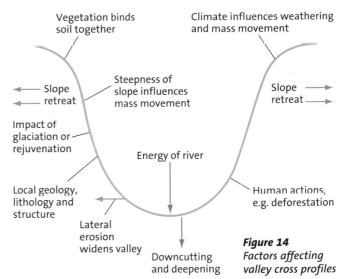

**Figure 14**
*Factors affecting valley cross profiles*

---

**7** Using case studies

## Question

(a) Explain the differences between weathering and mass movement.

(b) (i) Make a large copy of Figure 13. Identify and label the landforms that make up the valley sides.

(ii) Explain how two of these features have been formed.

## Guidance

Outline sketches are a useful way of showing information and helping understanding. Figure 14 is a summary of the factors shaping valley cross profiles.

---

# Long profiles

Landforms and processes change with distance downstream in a recognisable pattern, largely as a result of changing discharge. In the long term, river systems adjust to create a balance between energy available and energy used. This adjustment creates a characteristic downstream gradient called the **graded long profile** (Figure 15A).

However, this process is often interrupted by natural irregularities such as geology, resulting in waterfalls and lakes (Figure 15B).

The controlling factor in this system is the base level, which may be the altitude of a waterfall or confluence, the level of a lake or, ultimately, sea level. These interruptions alter the downstream gradient of the river, causing the river to change. Over time, waterfalls gradually disappear and lakes are filled in with sediment.

If sea level falls, this gives additional energy to the river system and erosion increases. This rejuvenation process not only causes the river to cut downwards more rapidly, but also increases headward erosion. The retreating 'head' of erosion

**Figure 15**
*Long profiles*

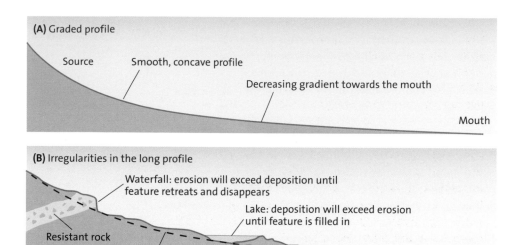

**(A) Graded profile**

Source

Smooth, concave profile

Decreasing gradient towards the mouth

Mouth

**(B) Irregularities in the long profile**

Waterfall: erosion will exceed deposition until
feature retreats and disappears

Lake: deposition will exceed erosion
until feature is filled in

Resistant rock

Eventual graded profile

Mouth

is marked by a sudden change in the river's gradient, called a **knickpoint**. Waterfalls often occur at these points; examples like Niagara Falls are really spectacular, cutting deep gorges as they retreat. Smaller, local earth movements, such as faults, can affect just one section of a river.

## Case study 4 — LANDFORMS AND PROCESSES ALONG THE RIVER LUNE

The River Greta is one of a number of left bank tributaries of the River Lune (Figure 16) which illustrate changes in river landforms, valley profiles and river long profiles. The Lune rises on the Howgill Fells, where the steep glaciated valleys give rise to turbulent streams, but the smaller Greta has its source in the limestone country on the western edge of the Yorkshire Dales National Park. Three locations along these rivers are worth closer examination:

**Figure 16**
*The Lune catchment*

Howgill Fells

Thornton Force

River Greta  **Burton in Lonsdale**

River Lune

River Wenning

**Morecambe**

Lune Floodplain

**Lancaster**

- Near its source above the village of Ingleton, the River Greta has features typical of the upper reaches of a river: a narrow valley and a dominance of erosion. Structural faults and rock types like Carboniferous limestone influence this river's landforms.
- Downstream, other river features begin to develop as the balance of fluvial processes changes.
- After reaching the main Lune valley, a wide floodplain develops as sediment is deposited and river channels migrate.

### Thornton Force

Thornton Force (Figure 17) is a 10 m high waterfall, 6 km from the source of the River Greta. This shows the importance of rock type and river processes in shaping upland river landforms.

*Figure 17* Thornton Force

Bob Hordern

- Upstream of the waterfall (the background of the photo), there are steep slopes, limestone scars, some soil creep, a V-shaped valley and interlocking spurs.
- Within the river channel, large angular boulders use up the stream's energy. Erosion occurs mainly in times of high discharge, when these boulders are 'rolled' downstream (**traction**).
- At the site of the waterfall, a horizontal layer of Carboniferous limestone called Great Scar forms the overhanging lip of the falls. Beneath this is a softer band of rock that is being eroded by water splash. Over time, this overhang is undercut and rocks fall into the plunge pool below. Hydraulic action gradually deepens this pool.
- The waterfall is gradually retreating upstream, creating a steep-sided gorge called Swilla Glen, noted for its geological sites and turbulent stream channel. The more effective process of corrasion uses material picked up by the stream to wear away the channel sides and bed of the river.

## 8 Question

Study Figure 17.
(a) Describe the landforms above and below the falls.
(b) Suggest why a waterfall may have occurred at this point along the river.
(c) Draw a side view of the waterfall and label the features and processes at work.

## Guidance

In examinations it is important to recognise and describe features. Diagrams of case study landforms quickly earn marks. Practise both skills in these questions.

### Burton in Lonsdale

The River Greta at Burton in Lonsdale (Figure 18) shows how changes in river channels and valleys can happen over small distances. Only 7 km downstream of Thornton Force, these changes are already apparent as the river and its valley are now wider, with different landforms in place.

- Within the channel, lateral erosion is now cutting into the bank on the right of the photo. This corrasion process is creating a river cliff as the **thalweg** (line of maximum velocity) swings to the outside of the meander. Near the inside bend, the water is slower and shallower, leading to deposition. The material left here on the point bar is still fairly large, but it is already less angular than it was upstream. Between the bar and the slip-off slope of the river bank there is short channel (or chute) through which water rushes when discharge increases.

*Rivers & Coasts*

Using case studies

■ Beyond the normal river channel, a small floodplain has formed where overbank discharge has deposited silt. The marked step at the foot of the woodland may be the remains of an earlier floodplain the river has cut into as it has gradually changed its course. River terraces like this are cut by downward erosion; river bluffs are reduced by lateral erosion.

**Figure 18**
*The features of a developing river meander*

Bob Hordern

## Question

(a) **Make a large copy of Figure 18. Identify and label the following river landforms on the sketch.**
   - river cliff
   - thalweg
   - point bar
   - slip-off slope
   - chute
   - floodplain
   - river terrace

(b) **Explain why the channel and valley features in Figure 18 look different from those in Figure 17.**

## Guidance

Using technical terms is another examination skill. In part **(b)** remember to compare both channel *and* valley.

### Lune floodplain

The lower Lune floodplain (Figure 19) is already 1 km wide below its confluence with the River Greta, and more changes are taking place regarding processes and landforms.
■ In flood conditions, which occur frequently (Figure 20), water spills onto the floodplain, depositing its suspended load as fine silt. Material adjacent to the river is coarser in calibre and is deposited before overbank velocities begin to decline.

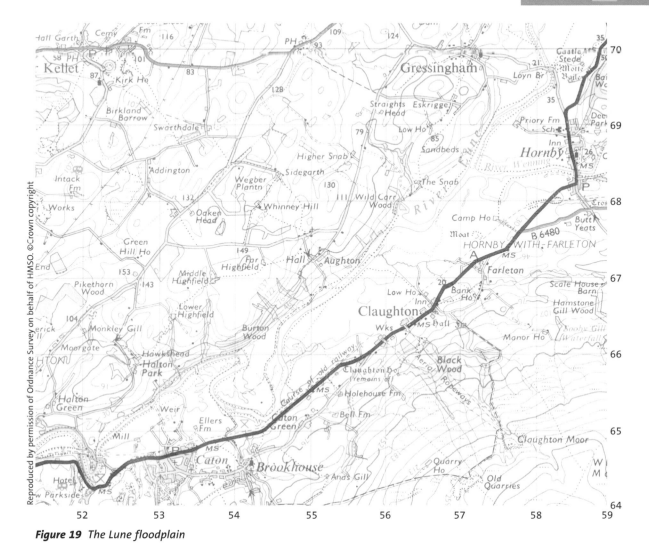

***Figure 19*** *The Lune floodplain*

***Figure 20*** *Flooding in the Lune floodplain, looking downstream and southwest towards Caton*

- Within the channel, deposition and erosion vary locally as a result of **helical** or **helicoidal flow**. This corkscrew movement of water in the channel has two effects: creating the sinuosity (curviness) of meanders and producing the alternating shallow riffle and deeper pools sections along the river bed.
- Braiding is common in times of low flow (Figure 21). The downstream gradient of the river here is low and there is a loss of competence (the ability to carry load). Coarser river gravels are deposited in the river by the shifting currents, creating temporary bars and islands. These are best seen at times of low flow.
- Meander migration is shown by sites of previous confluences and oxbow lakes. The largest meander between Caton Green and Burton Wood looks likely to be cut off in the future.
- The **wavelength** of River Lune meanders is slightly more than the standard '10 × river width' and there is little or no levée development.

**Figure 21**
*Braiding, riffles and point bars*

Bob Hordern

## 10 Using case studies

### Question

(a) Using Figure 19, identify the main river landforms, giving local names and grid references where possible.

(b) What evidence is there that river landforms are largely formed by deposition (see Figure 21)?

(c) Suggest how a high flood risk has influenced land use, settlement and transport.

### Guidance

For a closer look at this area, and to help you answer the questions, visit the following websites:

(i) www.multimap.com Search this site for Caton (choose option 1 for Caton, Lancashire). Choose the 1:25 000 scale. Pan northeast to view the meander and upstream. Click the camera icon to change the view to an aerial photo.

(ii) www.environment-agency.gov.uk In the 'Are you at risk of flooding?' dialogue box, enter the postcode LA2 9JA. Click on 'View map of these results' to see a map of the area. Change the scale to zoom in or out.

## Analysis of the long profile of Rivers Greta and Lune

The effects of interruptions and rejuvenation on the long profile of the two rivers (see Figure 22) are explained below.

1  The steep early course of the River Greta runs southwards on impermeable Yoredale rock (mostly peat-covered shales and grits).
2  The river then runs through Kingsdale, which was a lake at the end of the ice age trapped by a moraine. Now empty, the lake floor is being filled with deposition by the river (Figure 23).
3  The waterfall at Thornton Force (Figure 17) is the present limit of headward erosion (the moraine can be seen to the left of the photo).
4  Swilla Glen is the narrow gorge below the falls created by headward erosion.
5  Pecca Falls, within the gorge, is a knickpoint as the river responds to local earth movements caused by nearby Craven Fault.
6  The developing meander (Figure 18) near Burton in Lonsdale shows evidence of increasing lateral erosion.
7  The tributary Greta joins the main River Lune at its almost level floodplain.

**Figure 22**
*The long profile of the Rivers Greta and Lune*

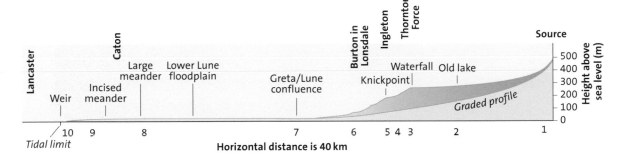

Horizontal distance is 40 km

Bob Hordern

**Figure 23**
*Kingsdale: a former glacial lake, with the river running towards the camera on the right*

8 Features linked to deposition, such as braiding (Figure 21) and the large meander near Burton Wood, appear. Floods are common in this area (Figure 20).
9 A fall in sea level in the past has caused the River Lune to cut down and create a narrow gorge at Halton (Figure 48), and an incised meander at Crook o' the Lune.
10 The artificial weirs upstream of Lancaster mark the limit of tides.

*This case study illustrates how, in each zone of a river, characteristic landforms and processes occur. As the balance between these processes changes, particularly downstream, the landscape changes. It also shows how general patterns, for example in the evolution of a river's long profile, can be interrupted by factors at specific locations. Valley cross profiles include the valley sides as well as the river channels.*

**11**

## Question
**Describe some of the ways in which either river long profiles or valley cross profiles are affected by geology and other factors.**

## Guidance
This is a good revision question to test your understanding of river landforms and processes. Use information from *Case studies 3* and *4* to illustrate your answer.

# Changing river channels

Detailed study of how river channels change, particularly downstream, can tell us much about how rivers work. It is also possible to investigate change in a number of stream variables, such as channel shape, velocity, discharge, load and gradient.

*Figure 24*
*Changing channel cross-sections*

## Channel shape

One way to view what happens in rivers is to understand how the channel continually adjusts to changes in inputs of water and sediment. River velocity and discharge are a product of the energy produced by gravity and the friction created by turbulence and channel friction. It has been suggested that in mountain streams, 90% of a steam's energy is used up by frictional drag and a further 5% is used to move the sediment load. Channel shape is therefore important in determining how efficient a stream's channel is. The term used is **hydraulic radius**.

$$\text{hydraulic radius} = \frac{\text{area of cross-section}}{\text{wetted perimeter}}$$

The higher the value, the more efficient the channel. The roughness of the stream bed also increases the wetted perimeter and so slows the velocity of the stream.

The shape and size of river channels change downstream as the width and depth increase. The characteristic V-shaped or other irregular cross-sections gradually give way to more rounded forms. In meandering channels, the helical flow is dragged towards the outer bends and the thalweg creates alternating asymmetrical cross-profiles. As illustrated in *Case study 4*, these profiles are closely linked to water velocity and the processes of erosion, transportation and deposition. Figure 24 summarises how channel cross-sections change downstream.

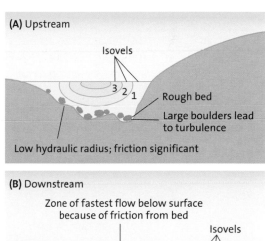

**(A)** Upstream

Isovels

3 2 1

Rough bed

Large boulders lead to turbulence

Low hydraulic radius; friction significant

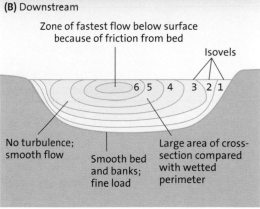

**(B)** Downstream

Zone of fastest flow below surface because of friction from bed

Isovels

6 5 4 3 2 1

No turbulence; smooth flow

Smooth bed and banks; fine load

Large area of cross-section compared with wetted perimeter

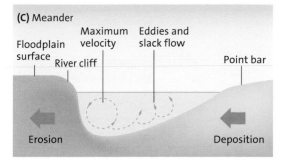

**(C)** Meander

Maximum velocity

Eddies and slack flow

Floodplain surface

River cliff

Point bar

Erosion

Deposition

**Question**

Using Figure 24, explain how the pattern of isovels in A and B changes with channel shape; compare the meander cross-section C with Figure 18 on page 16.

**Guidance**

Upstream, friction and low discharge dominate, while downstream the channel cross-section is more efficient and discharge is greater.

# Velocity and discharge

## Fieldwork

Velocity and discharge are two river variables that can be measured using fieldwork techniques. The strategy depends on the type and size of stream under investigation, the number of helpers and the equipment available. Having decided on a suitable stream and carried out a risk assessment, it is best to identify around ten sites (if you want to use a statistical test) where access is possible and measurements can be taken safely. Plan a sampling strategy that is either stratified or systematic. Where a detailed, small-scale focus is used, such as with a single meander, no sampling is required.

- Velocity can be recorded using a float and stopwatch, a flow meter or a salt dilution test. (To improve accuracy, surface velocity results should be multiplied by 0.85. This gives a more useful mean velocity).
- The area of cross-section can be calculated by multiplying the average depth by the average width.

An excellent river fieldwork website about the River Darent can be found on the GeoResources website at www.georesources.co.uk/darentintro.htm

## Secondary research

Changes in the quantity and quality of discharge, both downstream and over time, can also be researched using information from sources like the Environment Agency website (www.environment-agency.gov.uk). River discharge data in the UK are available from the National River Flow Archive website (www.nwl.ac.uk/ih/nrfa), which can be downloaded and used with spreadsheet programmes. These reports not only explain how discharge changes, but also explain what physical and human factors cause these changes.

*Case study 5*

## INVESTIGATING THE RIVER DERWENT

### Fieldwork

The River Derwent in Derbyshire rises on Howden Moor at around 575 m above sea level. Streams that drain this gritstone moorland are a good place to collect first-hand data about channel variables. This was done at points along related first-order, second-order and third-order streams (Figure 25). The measurements were taken during low flow and show changes in channel form, cross-section and other variables over a 1.5 km section of the river. Data about bankfull stage were also collected to calculate hydraulic radius and to help predict flood levels.

**Figure 25**
*Typical cross-sections and fieldwork sites*

**(A)** Bull Clough: first-order stream

The valley here is small and steep-sided, and has interlocking spurs and a cover of rough grassland. The channel is narrow and the downstream gradient is interrupted by mini-waterfalls and large rock debris.

**(B)** Cranberry Clough: second-order stream

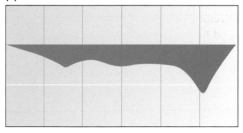

The valley is now widening and has a straighter course. Velocity and wetted perimeter have increased and downstream gradient has generally fallen.

**(C)** River Derwent: third-order stream

The Derwent valley is more open and the channel much wider. There has been a noticeable increase in discharge. Large boulders, moved in flood conditions, are stranded along the river bed.

Dot Kell and students

### Fieldwork results

The data from the 11 sites surveyed (Table 1) reveal a clear pattern in the different stream orders in this section of the river. Five variables — gradient, mean channel width, mean channel depth, normal stream discharge and mean velocity — were selected for analysis as these relate to one of the models used by geographers to help understand how river variables change downstream (Figure 26).

**Figure 26** *The Schumm model of how stream variables change downstream*

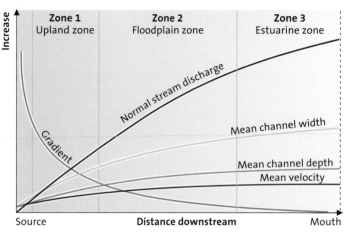

| Site name | Site number | Stream order | Gradient (degrees) | Discharge (cumecs) | Channel width (m) | Channel depth (cm) | Velocity (m s$^{-1}$) | Area of cross-section (m²) |
|---|---|---|---|---|---|---|---|---|
| Bull Clough | 1 | 1 | 60.0 | 0.17 | 2.3 | 28.0 | 0.27 | 0.64 |
| Bull Clough | 2 | 1 | 14.0 | 0.03 | 1.2 | 11.0 | 0.24 | 0.13 |
| Bull Clough | 3 | 1 | 10.0 | 0.01 | 1.2 | 26.0 | 0.05 | 0.31 |
| Cranberry Clough | 4 | 2 | 5.0 | 0.03 | 1.4 | 11.0 | 0.21 | 0.15 |
| Cranberry Clough | 5 | 2 | 8.0 | 0.14 | 2.3 | 13.7 | 0.45 | 0.31 |
| Cranberry Clough | 6 | 2 | 6.0 | 0.32 | 3.3 | 26.3 | 0.37 | 0.86 |
| Cranberry Clough | 7 | 2 | 0.0 | 0.10 | 2.6 | 13.5 | 0.29 | 0.35 |
| River Derwent | 8 | 3 | 4.0 | 0.18 | 2.6 | 22.7 | 0.30 | 0.59 |
| River Derwent | 9 | 3 | 1.5 | 0.13 | 3.2 | 8.4 | 0.50 | 0.27 |
| River Derwent | 10 | 3 | 6.0 | 0.69 | 7.0 | 20.9 | 0.47 | 1.46 |
| River Derwent | 11 | 3 | 10.0 | 1.34 | 5.4 | 26.2 | 0.94 | 1.42 |

**Table 1**
*Fieldwork results*

### Fieldwork analysis

It is not always easy to analyse data using a table, so the five variables have been graphed (Figure 27). This often clarifies the patterns found and allows explanation.

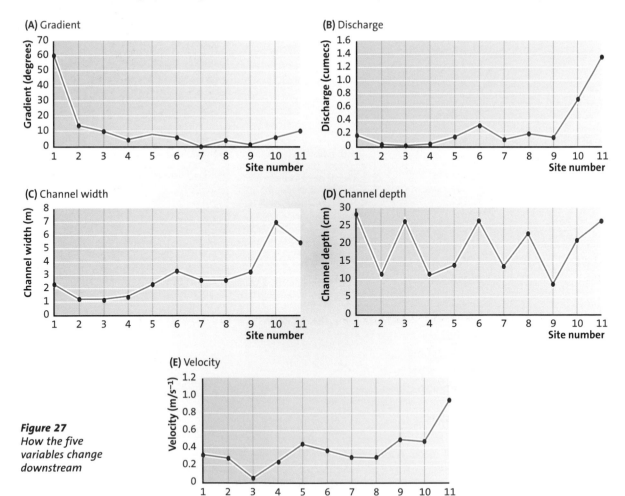

**Figure 27**
*How the five variables change downstream*

- Gradient: as expected, the gradient is initially very steep at 60°, but it soon falls to single figures. It almost flattens out as Cranberry Clough enters the River Derwent. Small, local variations may be due to the effects of geology: the Slippery Stones are large slabs of sandstone rock found in this area, which resist erosion.
- Discharge: this shows some variation but increases rapidly to 1.34 cumecs on reaching the River Derwent. This results in part from the higher volume of water produced as tributary streams join the main river. The stream velocity and hydraulic radius are increasing too as the channel becomes more efficient.

In conclusion, it seems the results were as expected. However, the model used was conceptual and so has no values or units to compare with real data. It also looks at a whole river, not just one section, and is probably derived from numerous observations and not based on only one day's readings.

## 13 Question

**(a)** Complete a written analysis of the remaining three variables: channel width, channel depth and velocity.

**(b)** Figure 26 is one model used by geographers to examine how stream variables change downstream. Research an alternative model (e.g. Bradshaw's model). Are there any differences?

## Guidance

**(a)** Read the analysis of gradient and discharge results to help you complete the answer.

**(b)** Bradshaw's model is in most post-16 textbooks.

*Using case studies*

## Secondary research

The 1200 km² Derbyshire Derwent catchment flows southwards for some 80 km to join the River Trent (Figure 28). Activities in the northern section are rural, focusing on recreation and farming. The middle section has a legacy of textile manufacture and water pollution. The lower course is more urbanised with large industries and a growing transport infrastructure.

The River Derwent Local Environment Action Plan (LEAP) gives useful background information and explains how and why discharge changes along the River Derwent. The *CAMS* (Catchment Abstraction Management Strategies) publications provide an update on catchment issues. The National River Flow Archive website at **www.nwl.ac.uk/ih/nrfa** provides a more detailed picture of what happens to the Derwent's discharge downstream, with data from flow gauges that continually measure discharge at a number of stations in the catchment. Table 2 summarises information from all these resources.

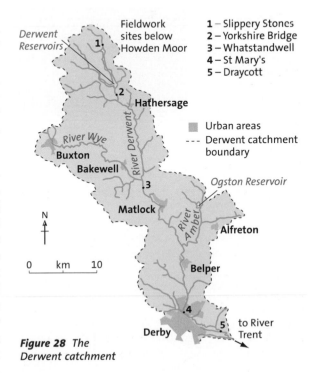

**Figure 28** The Derwent catchment

| Flow station | Distance downstream (km) | Mean annual discharge (cumecs) | Grid reference OS 43 (SK) | Notes |
|---|---|---|---|---|
| 1 Slippery Stones | 6 | 0.59 | 169 951 | The gritstone moorlands near the river's source deliver water into the three Derwent reservoirs |
| 2 Yorkshire Bridge | 18 | 2.10 | 198 851 | Below the reservoirs, with water having been abstracted to supply Sheffield, Nottingham and Derby |
| 3 Whatstandwell | 53 | 16.02 | 331 545 | Below the confluence with the River Wye, which drains the limestone aquifer supplying the baseflow of the Derwent |
| 4 Derby, St Mary's | 68 | 17.48 | 355 368 | Balance between water abstraction and returning effluent from Derby; local sandstone aquifers provide valuable support as rainfall levels here are below 700 mm per year |
| 5 Draycott | 77 | 20.08 | 443 327 | Quick runoff from urban drains and impermeability; water demand is high as the majority of the catchment's 375 000 residents are concentrated here |

**Table 2**
*Summary of five flow stations along the River Derwent*

**Figure 29**
*Ladybower Reservoir — one of the Derwent reservoirs*

## 14 Question

(a) Use the data in Table 2 to draw a line graph showing discharge against distance downstream. Add a best fit line.

(b) Comment on:
  (i) the general pattern shown by the graph
  (ii) why Yorkshire Bridge (station 2) has a relatively low discharge
  (iii) why Whatstandwell (station 3) has a relatively high discharge

### Guidance

(a) ■ Make the *x* (horizontal) axis of your graph your distance downstream line.
  ■ The *y* (vertical) axis should be used for the dependent variable, i.e. discharge.
  ■ Use an appropriate scale and remember to label both axes.

(b) Refer to the notes in Table 2 and the map (Figure 28) when answering this question.

Research also shows the importance of the Derwent reservoirs, built in the upper catchment in the early part of the twentieth century. They provide valuable water supplies for nearby cities and recreation opportunities in the Peak District National Park. However, they also have a marked impact on the River Derwent's discharge (Figure 30).

**(A)** Flow station 1 (Slippery Stones): above the reservoirs

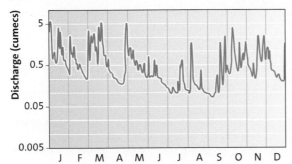

The hydrograph data above the reservoirs at Slippery Stones are flashy, with many peaks and troughs throughout the year. This is consistent with an upstream location where rainfall is high and varied, with a minimum in the summer months. Lag times appear short as water quickly runs off the steep, impermeable slopes. The mean flow is 0.7 cumecs.

**(B)** Flow station 2 (Yorkshire Bridge): below the reservoirs

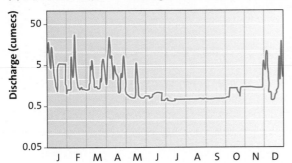

Note the larger range of values on the discharge axis. At Yorkshire Bridge, downstream of the reservoirs the winter and spring peaks reflect events upstream (A), as the reservoir fills and discharge is passed downstream. By May, rainfall is less and evaporation reduces storage levels. The flow is then regulated by the dams. In Autumn, leaf fall and cooler temperatures allow more water to run off. The mean flow is 2.31 cumecs.

**Figure 30**
*Typical annual hydrographs for flow stations 1 and 2*

*This case study complements ideas about river landforms and processes in Case study 4. It shows how fieldwork and secondary research can be used to investigate rivers. It also examines how channel variables, such as cross-profile, velocity and discharge, change downstream.*

# Velocity and load

Rivers receive 90% of their load from the denudation of their surrounding catchments; the rest is eroded from within their channels. Once in the river, the relationship between the velocity of the river and this material determines what happens in the channel. For this topic, you should try to learn and use the following key terms:

- **entrainment**: the process of picking up and transporting the load
- **calibre**: the size of individual particles
- **competence**: the maximum size of particle that can be entrained
- **capacity**: the total mass of load that can be moved by a river

The relationship between river velocity and the calibre of the load can be shown graphically. This is called the Hjulström curve (Figure 31).

- The *x*-axis shows particle size from fine clay (less than 0.4 mm diameter) to large boulders (more than 256 mm diameter). The *y*-axis shows velocity (usually with a logarithmic scale).
- The graph shows three areas — erosion, transportation and deposition — bounded by lines which are the pick-up and drop curves. Between these two boundaries, particles remain in transport (particularly the smaller ones in suspension).
- At first glance, it seems that high velocities lead to erosion, fine particles are carried most easily and larger particles are deposited first.

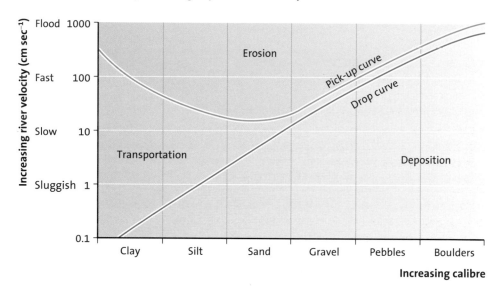

**Figure 31** *The Hjulström curve*

**Question**

Look at Figure 31.

(a) A stream has a baseflow of 10 cm per second (cm s$^{-1}$). What is the largest size of particle it can transport?

(b) During a storm, the velocity increases to 100 cm s$^{-1}$. What is the largest particle it can now erode?

(c) This is followed by a period of drought and the river's velocity falls to 1 cm s$^{-1}$. Describe and explain what happens to the load during this transportation and deposition phase.

(d) Copy and complete these sentences:
   (i)  The pick-up curve is always above the drop curve. This means it requires higher...
   (ii) On the left of the graph, the pick-up curve rises, meaning finer particles...
   (iii) These particles are demonstrating cohesion, which means...

**Guidance**

Place a ruler horizontally on the graph to help you read values accurately. For example, in part **(a)** position the ruler through the 10 cm s$^{-1}$ point. Sand should be the calibre being transported. Repeat this strategy.

*Contemporary Case Studies*

Recent research suggests that it is not so much the velocity or discharge of a stream that determines its competence and capacity, but how the water actually flows in the channel. Laminar, turbulent and helical movements of water may well be the key to new theories about entrainment.

However, data from the Santa Clara River in California (Table 3) show that there are clear links between stream discharge and the amount and calibre of the load.

- The first row of data show that when discharge is very low (13 cumecs), so is the amount of sediment load (232 tonnes per day). In addition, most of this load is fine (73% clay).
- In contrast, the second row shows how a very high discharge (4420 cumecs) can carry a large amount of sediment (736 000 tonnes per day). The calibre of the load is now more balanced, with coarser sands and gravels also being transported downstream.

| Discharge (cumecs) | Sediment load (tonnes day$^{-1}$) | Particle size (%) | | |
|---|---|---|---|---|
| | | Clay | Silt | Sand and gravel |
| 13 | 232 | 73 | 27 | 0 |
| 4420 | 736 000 | 26 | 34 | 40 |
| 1860 | 756 000 | 30 | 33 | 37 |
| 1645 | 534 000 | 33 | 41 | 26 |
| 1830 | 321 000 | 36 | 35 | 29 |
| 923 | 92 000 | 35 | 37 | 28 |
| 40 | 323 | 72 | 27 | 1 |
| 870 | 105 000 | 33 | 41 | 26 |
| 350 | 103 000 | 26 | 38 | 36 |
| 6250 | 1 196 000 | 23 | 37 | 40 |
| 15 | 172 | 57 | 40 | 3 |
| 1350 | 120 000 | 37 | 40 | 23 |
| 1980 | 253 000 | 55 | 41 | 4 |
| 3200 | 513 000 | 25 | 34 | 41 |

**Table 3**
*Data from the Santa Clara River*

Discharge and velocity are not the only variables influencing load. The supply of material and the nature of the stream are also important. Not all load is carried in suspension. In limestone areas (*Case study 1*), **solution** may be the main process at work. Bedload material is also moved along in traction. Turbulence may reduce the efficiency of the channel to move water, but on the other hand it actually helps dislodge and move sediment particles. These may then be slid, rolled or bounced (**saltation**) along the channel.

Deposition also occurs within the channel and beyond. In meandering streams, this takes place on point bars, while braiding occurs where gradient or flow decreases. In river mouths, deltas and mudflats are common depositional features. Again, these features are not solely about falling water velocities. Other factors are also involved, such as salinity, which causes fine particles to **flocculate** or stick together. Ecosystems such as saltmarsh and plants like spartina also help trap sediment. Wetland, beaches and sand dunes are the recipients of much of this progressively fined material.

## 16 Question

**Using case studies**

Study Table 3.
(a) Do other rows of data show similar trends? What are your conclusions?
(b) How could you use a suitable statistical technique to examine these relationships in the Santa Clara River?

## Guidance

In part **(b)**, a scattergraph and best-fit line are useful, but you could use Spearman's Rank and then test for significance.

# Rivers and people

Rivers have always presented people with opportunities and challenges. From earliest times, people have exploited rivers for water supply and transport. Natural flooding provided valuable silt for farmers, irrigation allowed the production of surplus food resources and river navigation brought increased trade. Cities grew along the banks of rivers as roads and railways followed their courses. Later, industry used the water for mechanical power and the generation of electricity. In modern times, recreation also competes for waterside sites. Riversides are desirable locations.

However, this exploitation is not without challenges. Physical factors and natural hazards bring the risks of flooding, low flow, sedimentation and changing river channels. These are all disadvantages of using floodplain and estuary locations.

Conflict over the use of rivers and their catchments is inevitable. Economic demands and environmental concerns are not always easily reconciled. This is particularly true of large rivers, where water and power supply issues may have international implications.

People have not always managed these areas wisely. Interference with natural processes, poor land management and inappropriate development can make matters worse. Many flood disasters are caused by human intervention, and part of their impacts can be put down to poor management. Some problems, such as pollution, are caused directly by human activity; other interventions, such as building dams, may have unforeseen consequences. Impacts on linked coastal systems can also be detrimental to the environment.

## Opportunities and challenges

Rivers present both opportunities and challenges. For example, regional and national economic benefits from the Three Gorges Dam above Wuhan in China must be weighed against local environmental and social costs. Is the loss of important habitats and forced human migration worth the gain in industrial development and flood protection? These issues require an understanding of how rivers influence people's lives and how human activities can impact on river environments. Many of these issues are interrelated:

- Flooding brings risks to life and property, loss of food supplies and ill health.
- Low flow causes ecological loss, water supply problems and sedimentation issues.

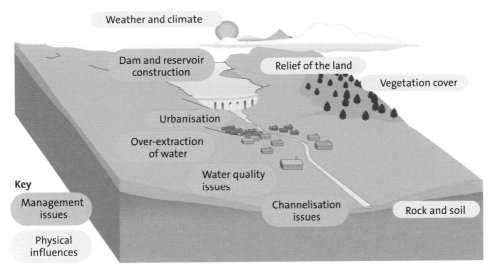

**Figure 32**
*Interrelationships between physical influences and management issues in a drainage basin*

- Water resource issues include extraction, abstraction and effluent problems.
- Erosion brings loss of soils, land and property.
- Sedimentation impacts on flooding, water quality and navigation.
- Channel migration causes property damage and navigation difficulties.
- Pollution leads to ecosystem damage, water supply issues and loss of amenities.

One way of summarising people–environment interrelationships in river catchments is shown in Figure 32, which identifies the physical factors at work as well as the human activities and management taking place.

## THE SAN GABRIEL RIVER

*Case study* **6**

The 90 km long San Gabriel River in California illustrates many of the interrelationships between people and rivers explored in this part. Its catchment covers an area of 1837 km² between the southern Sierra Nevada Mountains and the Pacific Ocean.

The San Gabriel River catchment can be researched in more detail using information on the County of Los Angeles Department of Public Works website (**www.ladpw.org/wrd/publication/system**).

Flood, drought and sediment challenges in the San Gabriel catchment arise from the physical factors, natural hazards and human activities.

The climate is semi-arid, with annual rainfall levels lower than anywhere in the UK. During the summer months there is prolonged drought and the ground is baked by the high insolation. Any rainfall either runs off rapidly or evaporates. Most rainfall occurs in winter and it is then that discharge levels can rise dramatically in response to intense storms. Inland, the annual figures in California are much higher than on the coast. A water (soil moisture) budget graph is a useful way to study the interplay of climate, hydrology, soils and vegetation (Figure 33). This graph shows how precipitation and temperature:

- change throughout the year
- affect potential evapotranspiration
- affect soil moisture throughout the year
- are influenced by vegetation via transpiration and osmosis

**Figure 33**
*A soil moisture budget graph for California*

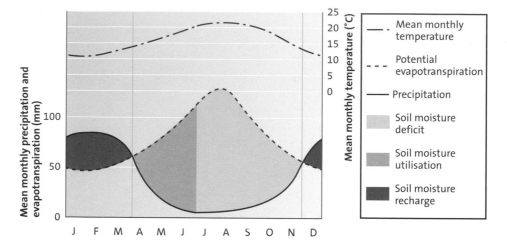

Vegetation in upland areas is sparse and mostly scrub, with scattered trees — a type of vegetation known as chaparral. Many plants are xerophytic and in the summer drought fires are commonplace. This reduces the ability of the vegetation to intercept rainfall in the autumn. Figure 34 shows typical sage bushes and a dry streambed.

**Figure 34** *Typical scrub vegetation*

Bob Hordern

**17** *Using case studies*

## Question

Study Figure 33.
(a) Describe how the soil moisture budget conditions change throughout the year.
(b) Suggest why these changes happen.

## Guidance

Focus on the three different coloured areas of the graph. Key factors are temperature, precipitation, interception and evapotranspiration.

Mountains in the San Gabriel range rise to over 2000 m and have steep, west-facing slopes. Figure 35 illustrates the sudden change in gradient. Slopes are dissected by a vast network of canyons that remain dry for much of the year but can become raging torrents in the winter rainfall.

Rocks and soils are easily weathered because of the high temperatures. The thin veneer of material that results is easily eroded during storms, when gullying and mudslides occur along canyon walls. Flash floods have large sediment loads. Rock types in the basin are both impermeable and permeable and this has been used to advantage by engineers to surface store water and recharge aquifers.

Rapid urbanisation — the Los Angeles area has 15 million residents — has had two hydrological impacts on the catchment basin:
■ It has created an enormous demand for water.
■ It has led to increasing floods, most notably in 1914, 1938 and 1969.

By the 1950s, much of the area around the city of San Bernardino had also become urbanised, with affluent Californians building homes within the lower canyons of the San Gabriel mountains. The effects of this urbanisation can be seen in Figure 36, which shows the different responses to a storm of two neighbouring streams: one through farmland and the other in a built-up area.

Bob Hordern

*Figure 35*
*Steep slopes give way to level lowland*

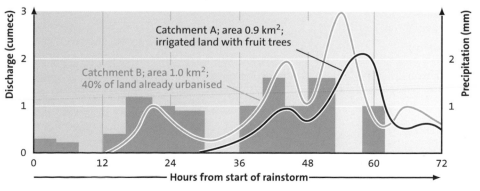

*Figure 36*
*The effects of urbanisation on two Californian streams*

## Question

Study Figure 36.

(a) Compare the responses of the two streams to the rainstorm.

(b) Suggest reasons for any differences.

## Guidance

In these comparisons the usual factors are related to urbanisation — drains, culverts, removal of vegetation and impermeable surfaces.

Opportunities for flood and sediment control, water supply and recreation have been taken using dam and reservoir construction, channelisation, water abstraction and recharge.

Dam construction was seen as the foremost solution to water management issues in California. Los Angeles is served by a network of dams and channels designed to control discharge and sediment levels within the catchment and to supply water to its suburban areas. Features of the present San Gabriel scheme are given below and shown in Figure 37.

- The Cogswell and San Gabriel dams in the headwaters regulate flood water, sending it to permeable spreading grounds to recharge aquifers downstream.
- The Morris Dam in the canyon section further regulates discharge. The San Gabriel Debris basin collects bedload sediment from a tributary. This is later excavated for use as aggregate.

- The Santa Fe Dam, sited where the river enters the lowland, is usually dry, but its gates can be closed when floods threaten along the San Bernardino Freeway.
- The Whittier Narrows Dam in the lowland area has a crucial role, storing and diverting water into either the San Gabriel or Los Angeles rivers. In dry periods, this 5 km-long dam allows recreation (parks, golf courses, trails and fish ponds) within its 976 hectares.
- The Rio Hondo (linking to the Los Angeles River) and San Gabriel river then carry water away in a southwest direction to the coast, across the Montebello Forebay area.

Channelisation is used to deal with the wide, shallow, meandering rivers that form where canyons have opened out onto the lowland. These concrete water 'highways' were built to prevent rivers changing course through growing urban areas. Their smooth, enlarged cross-sections were designed to increase water velocity and carry floodwater quickly to the sea. Many channels remain dry in the summer.

**Figure 37**
*The San Gabriel water conservation scheme*

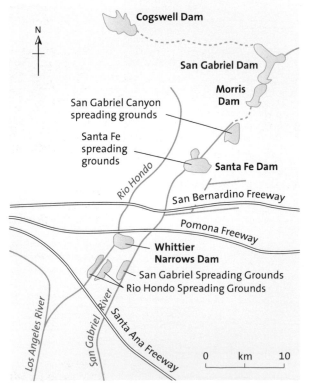

Water abstraction and recharge into downstream aquifers in past years was seen as vital given the massive demand in the area. Today, 60% of the water required in Los Angeles is brought great distances from sources such as the Parker Dam on the Colorado. The pressure that results from the artificial storage of fresh water in permeable rocks underground has the added benefit of preventing sea water from contaminating groundwater supplies.

*This case study shows how physical factors such as climate, and hazards such as flooding, pose challenges for people. It also examines how activities like urbanisation create challenges and opportunities for river management.*

# Flood hazards

River flooding is a natural event that frequently becomes a major hazard to people. Worldwide it is estimated to kill on average 25 000 people each year and affect a further 500 million. In addition, it damages buildings, infrastructure and livelihoods.

The causes and severity of floods relate to the precipitation involved and the character of the location in which they happen. The conditions that favour flooding are summarised in Figure 38. Notice how they may result from prolonged rainfall, flash flooding or rapid snow thaw, and have physical, human and other causes.

It is important to separate these causes of flooding from the effects they have on people's lives, and to understand how their scale and duration can differ. The effects of flooding can be direct and immediate — as in the case of drowning and property damage — or occur indirectly and after some time as a result of subsequent food and health issues.

Like all natural hazards, flooding has a frequency and a magnitude. One way to express this is in terms of the flood return period (in years) and the areas likely to be flooded by these events. This is shown on Figure 39, where floods may reach the edge of the floodplain only once in 500 years.

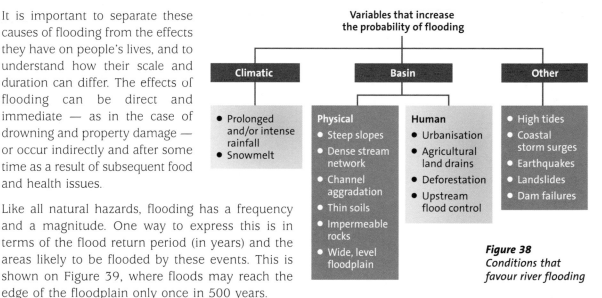

**Figure 38**
*Conditions that favour river flooding*

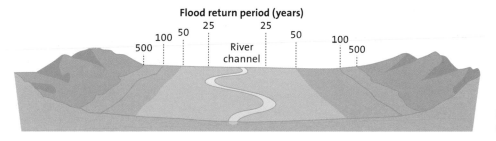

**Figure 39**
*Flood return periods*

## A FLOOD COMPARISON OF BOSCASTLE AND BANGLADESH

In August 2004, on opposite sides of the world, two flood events were taking place: one in a small village on the Cornish coast, the other across much of Southeast Asia.

### Boscastle

On 16 August, an estimated 310 million gallons of rainfall fell over a small river catchment (20 km²) in north Cornwall. This localised flash flood took no lives but swept away the homes, businesses and vehicles of almost 1000 people. The financial cost to this coastal tourist village was £62 million, based on insurance claims and loss of trade. Despite the lack of any warning, the emergency services arrived quickly.

Meteorologists blamed these events on a storm created by rapid uplift and convergence as moist air rose overland. Calculations suggested that perhaps 70 mm of rain (more than 1 month's average) fell in just 2 hours following 2 weeks of high rainfall. Geographers took a wider view by identifying other factors, such as the catchment's many steep slopes on impermeable rocks (slate) and the siting of the village at the confluence of three streams: the Valency, Jordan and Paradise. The temporary landslides — which first dammed but then later released a 3 m floodwave — and the unfortunate coincidence of a high tide, were also apparent. For local people and journalists, the floods were a reminder of the similar but more tragic Lynmouth disaster exactly 52 years earlier.

### Bangladesh

Between mid-July and mid-August, there was widespread flooding in northern India and Bangladesh. This recurring event is fed by rivers draining a combined catchment area of 1.5 million km². The death toll was around 900 people and the homes, food supplies and livelihoods of a further 20 million people were affected. Two thirds of Bangladesh was flooded and damage was estimated at almost £4 million. Some food and emergency supplies were donated by other countries, though disease and health risks were a considerable concern.

The monsoon season regularly brings heavy rainfall between June and September, but predicting its intensity and distribution is not easy. From a wider perspective, the country sits astride the delta of the Brahmaputra, Ganges and Meghna rivers, and population pressures encourage more people to settle in flood-prone areas. Urbanisation and development are increasing runoff on the floodplains. With only 10% of these rivers inside Bangladesh's borders, deforestation in Nepal and river management in India are not under its control. This situation is not new — similar events occurred in 1974, 1987, 1988 and 1998.

**19**

### Question

(a) **Identify similarities and differences between the causes of floods in Boscastle and Bangladesh.**

(b) **Both examples focus on the negative aspects of flooding. Suggest some positive impacts.**

### Guidance

Benefits include soil nutrients, aquaculture, and flushing.

# SHREWSBURY: LIVING WITH FLOODS

The flood risk along the River Severn affects people and businesses in towns such as Shrewsbury. Floods are a frequent occurrence, disrupting transport and businesses as well as residents. The causes are found upstream, with heavy rainfall on steep slopes leading to high runoff, followed by a number of river confluences. The town itself is built around an incised meander, with a relatively small area of floodplain.

Because of early flood warnings, flooding results in minimal social impacts, but economic impacts include property damage, additional staffing costs and loss of trade. Of particular interest is the influence of flood risk on building and land use decisions, which can be analysed in the aerial photo of Shrewsbury (Figure 40).

- Land along the river corridor consists of parks, school playing fields and tennis courts, car parks, a showground and a football ground. This open land may be damaged by flooding but it can recover relatively quickly. The river banks are planted with trees. In future, development within this 'most at risk' zone can be avoided.
- Beyond this immediate floodplain area, slightly higher ground protects property from flooding except where roads descend to the riverbanks. A mix of residential and commercial property is found here in the core of the meander.
- The highest areas approaching the neck of the meander mark the historic site of the town. Here the main shopping centres and offices are not at risk.

**Figure 40**
*An aerial photo of Shrewsbury*

Aerofilms

- The road network and bridges are largely built on the higher ground, as is the railway station beyond the meander neck.
- Outside the town centre, property is partly protected by the river cliffs that are found on the outside of what is a distinctive incised meander.

For further details, the September 1999 (Vol. 13, No. 1) and March 2000 (Vol. 13, No. 4) issues of *Geography Review* contain information on the flood events of October 1998. Photos of the 2004 flood are available on the Shrewsbury Guide website (**www. shrewsbury-guide.co.uk**). Click on 'Tourist guide', then follow the link to the photo gallery.

# River pollution

Pollution in rivers exists because of poor process control or poor environmental management. This is a human phenomenon. Major incidents make headlines, but it is large numbers of small events that together create environmental problems. Two types of source can be identified:
- **Point sources** of pollution, such as sewage outfalls, occur at specific sites and enter the river directly.
- **Diffuse sources**, like agricultural nitrates, spread their effects over a wide area, often in runoff or via groundwater.

It should not be forgotten that most emissions into rivers are legal, covered by licences and regularly monitored by the Environment Agency. Pollution can be detected by a variety of tests, ranging from watch scores looking at the effects on indicator organisms to measurements of biological oxygen demand (BOD).

Three of the most significant types of river pollution are caused by sewage works, agricultural activities and industrial accidents (Figure 41).

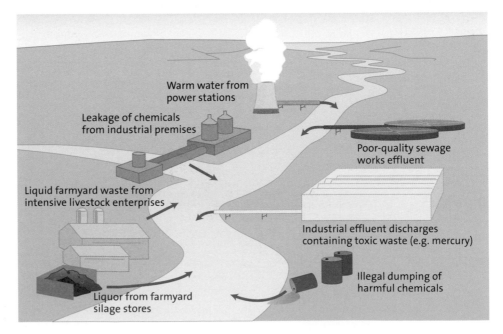

**Figure 41**
*Common causes of river pollution*

- Sewage from treatment works always has some effect on rivers, but it is when concentrations of bacteria develop in organic waste that problems develop. Bacteria use up oxygen in the water, threatening existing organisms.
- Farm pollution in the form of nitrates from artificial fertilisers or natural slurry can affect a wide area. Together with phosphates, they create algal blooms, leading to nutrient overload and eutrophication. Fish are particularly affected by this process, but nitrates threaten human health too.
- Chemical spillage can have rapid and devastating impacts, as the infamous Sandoz accident in Basle, Switzerland, demonstrated in 1986. In this case, 30 tonnes of pesticides, mercury and toxic waste entered the River Rhine, killing microorganisms, fish and birds in four countries.

## INVESTIGATING POLLUTION IN THE RIVER DARWEN

*Case study* 9

The River Darwen begins on open moorland before flowing north through Darwen to Blackburn where it is met by the River Blakewater. From there it flows westwards, through farmland to join the River Ribble before it enters Preston. From there it flows into the Irish Sea (Figure 42).

The valley has an industrial legacy: it was the scene of cotton weaving in the past, and in 1907 there were still 57 weaving mills and 8 spinning mills along its course. Today, a few mills remain, mostly making paper. Other sources of pollution are farms,

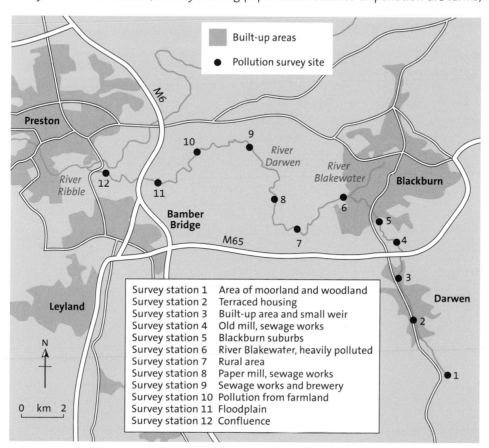

| Survey station 1 | Area of moorland and woodland |
| Survey station 2 | Terraced housing |
| Survey station 3 | Built-up area and small weir |
| Survey station 4 | Old mill, sewage works |
| Survey station 5 | Blackburn suburbs |
| Survey station 6 | River Blakewater, heavily polluted |
| Survey station 7 | Rural area |
| Survey station 8 | Paper mill, sewage works |
| Survey station 9 | Sewage works and brewery |
| Survey station 10 | Pollution from farmland |
| Survey station 11 | Floodplain |
| Survey station 12 | Confluence |

*Figure 42*
Map of the River
Darwen, Lancashire

which may cause nitrates to enter the river, a large brewery and various sewage works serving the urban areas of Darwen and Blackburn. Runoff and storm drains increase pollution risks in urban areas and litter is a problem in some locations.

An investigation of the River Darwin could include:

- fieldwork. Carry out water quality surveys (e.g. nitrate and phosphate content) and watch scores based on indicator fauna in the water.
- research. Use the Envioronment Agency website at **www.environment-agency.gov.uk**. Follow the link to 'Water quality' and then 'What's in your backyard?' In the dialogue box, enter the postcode PR5 0UY to see the data for the sewage works above survey station 10. Choose 'view maps of results' in the rivers section and a map of part of the River Darwen will appear. Carefully click once on the centre of the map. This should reveal the typical land uses affecting the river quality at station 10.

## A student's investigation

### *Fieldwork aims*

To what extent is the River Darwen polluted? What are the sources of this pollution? How does this affect organisms in the river?

### *Hypotheses*

Pollution is most likely to be caused by point sources. High levels of pollution may lead to limited diversity of fauna (high chemical content results in low watch scores).

### *Methodology*

Carry out sampling at 12 sites to allow collection of statistical data:

- Collect short-term 'snapshot' data from a litre sample of river water, tested for chemicals using a colorimeter and probe.
- Carry out a medium-term 'visual pollution' survey (more points suggests a worse environment).
- Carry out longer-term watch scores using kick sampling and organism survey (more organisms suggests cleaner water and more oxygen).

Alternative tests include BOD, Trent Biotic Index, RIVPACS etc.

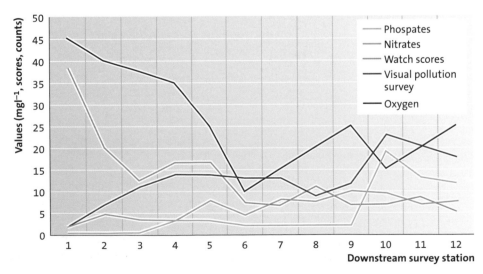

**Figure 43**
*Survey results*

*Data summary*
- Phosphates: increase from 4 onwards; jump to maximum at 10
- Nitrates: increase from 5 onwards; maximum at 9
- Watch scores: count of organisms falls; small improvement at 8
- Visual pollution survey: rising trend of penalty points; fall at 8; maximum at 10
- Oxygen: falling trend to 6; improvement from 7–9; sudden fall at 10

## 20 Question

**(a) Briefly summarise the main results of the River Darwen investigation.**
**(b) How does land use information explain the pattern of results?**
**(c) How far do the results allow the student to answer her aims and hypotheses?**
**(d) Suggest what the student might have written in her evaluation.**

## Guidance

This is a brief activity, so be concise. Land use is shown in Figure 42. An evaluation would consider methodology, reliability of results and the validity of the conclusions. The research task using the website should enable you to give clear answers.

# River conflicts

Rivers are valuable commodities, so it is not surprising that there is often much argument about how they are used and who benefits most from their exploitation. The big issues are economic, including water supply, flood control, hydroelectricity (HEP) and navigation. Politics are also involved, with questions of how resources are controlled and the environment protected. Many large rivers basins separate states and cross international boundaries.

At a local scale, issues arise when individuals or groups view development in different ways. The Environment Agency's Local Environment Action Plan (LEAP) initiatives try to identify the main players involved and plan how best to manage the river in question.

On a larger scale, major river systems are shared between states or countries, so agreement on use is vital. As demand grows in the future, arguments over rivers and water supplies may eclipse even those over land or oil resources. Some examples of disputes over the use of large rivers are outlined below:

- The Colorado compact was agreed in 1922, although the final details were not settled fully until 1963. Today, California's increasing demands, native peoples' claims about land rights and growing acknowledgement of the 'poor deal' for Mexico all threaten this agreement.
- The River Nile provides water for ten countries along its 6400 km journey. The present treaty, set up in 1929, favours Egypt and guarantees it two thirds of the river's water from the Aswan Dam. For the countries upstream, which have problems with drought and deforestation, this agreement is unfair and out of date, particularly as the population living in the Nile basin is expected to double in the

next 25 years. Sudan needs flood controls, Ethiopia wants reservoirs (dams) and Tanzania needs a water pipeline from Lake Victoria.

- In 1990, the annual flow of the Euphrates River was reduced from 30 km$^3$ to 16 km$^3$ when Turkey completed the Ataturk dam construction programme. Downstream, Syria has also built dams, but the reduced flow from Turkey is threatening its HEP supplies and irrigation projects at Tabaqa. By the time water reaches Iraq, the potential is almost gone and plans to use water from the tributary River Tigris have been affected by war.

## Case study 10  CONFLICT OVER THE MEKONG DAMS

**Figure 44**
**The Mekong dams**

The Mekong River flows through five countries from China to Vietnam. Dam-building on the Chinese section of the river, a disastrous drought and a growing population promise a difficult future for the region (Figure 44).

Upstream, China's new dam at Manwan provides HEP and water, allowing the development of Yunnan province. The second dam at Dachaoshan will be completed by 2012. More worrying still are the eight further dams planned. The flow of the river in Laos and Cambodia has been reduced considerably and fish stocks, which support 1.5 million people, are declining rapidly.

Downstream, Laos and Thailand also have plans for dams. In Laos, the US$1.1 billion Nam Theun Dam will displace over 5000 people and may bring enormous debts. In just 20 years, the Mekong River could be changed from being a relatively untouched river to one of the most dammed in the world, with more than 100 dams, diversions and irrigation projects. Rice production throughout the region is threatened by lowering water levels, and some scientists claim the proposed Sambor Dam above Phnom Penh would be an ecological disaster.

| | Conservation | Thermal power | Sewage treatment | Navigation | Industrial activities | Fishing | Farming | Recreation | HEP |
|---|---|---|---|---|---|---|---|---|---|
| Water supplies | | | | | | | | | |
| HEP | | | | | | | | | |
| Recreation | | | | | | ✔ | | | |
| Farming | | | | | | | | | |
| Fishing | ✔ | | | | | | | | |
| Industrial activities | | ✔ | | | | | | | |
| Navigation | | | | | | | | | |
| Sewage treatment | | | | | | | | | |
| Thermal power | ✗ | | | | | | | | |

**Figure 45**
**A conflict matrix**

**Question**

(a) Complete Figure 45 to identify some likely conflicting or compatible uses of a river.

(b) Use the information in Case study 10 to explain how today's gains can become tomorrow's losses.

**Guidance**

(a) Use ✘ for conflicts and ✔ for compatibility. You need not fill in all the boxes.

(b) Short-term economic benefits may give rise to long-term environmental costs.

# Mega-dams

Interfering with river systems can have negative and positive outcomes. These outcomes are seen most obviously in large reservoir and dam-construction projects.

The advantages and disadvantages of such schemes are summarised in Figure 46. China's Three Gorges Dam and Egypt's Aswan Dam show remarkable similarities to this model (Table 4).

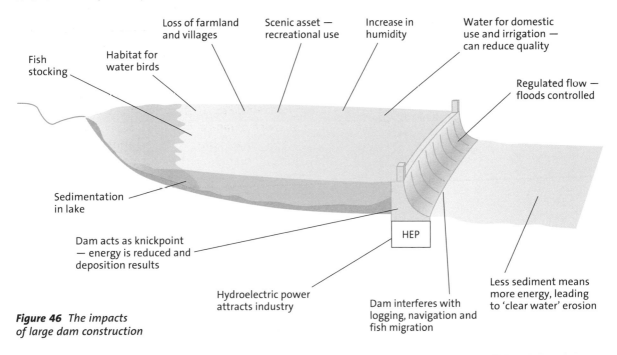

*Figure 46 The impacts of large dam construction*

Even the apparently successful San Gabriel water conservation scheme (*Case study 6*) is not entirely free from negative consequences. This scheme prevents floods, controls channel migration, provides water and creates recreation facilities, but there are some negative consequences, especially for the coastline:

- Annual discharge at the mouth of the San Gabriel River is now only 2% of the runoff upstream (although water enters the sea via the Los Angeles River).

- Annual sediment output at the coast is estimated to have been reduced by one third: perhaps some 601 000 m$^3$ is intercepted by the various dams. Total interception by all dams throughout the whole west coast region is reportedly 4.2 million m$^3$ of sediment.
- Experts suggest that 40% of this sediment is sand, so the impact on California's beaches and coastal management will be considerable. The annual loss is the equivalent of a beach 100 m wide, 10 m deep and 1.7 km long.

**Table 4**
*Advantages and disadvantages of two dams*

| Mega-dam | Year of completion | Advantages | Disadvantages |
|---|---|---|---|
| Aswan (Nile) | 1963 | • Flood control<br>• Increased food production<br>• Higher yields in cash crops<br>• HEP doubled<br>• River navigable<br>• Fishing on lake (better diet)<br>• New settlements and tourism | • Delta retreating<br>• Sea fishing poorer<br>• Salinisation of fields<br>• Bilharzia (disease carried by water snails)<br>• Less silt (collects in lake), so more fertiliser needed<br>• 20% loss of water to evaporation/seepage<br>• Heritage lost<br>• Some forced migration |
| Three Gorges (Yangtze) | 2009 | • Electricity output equal to 15 nuclear power stations<br>• Less reliance on coal<br>• Water supply for Shanghai (13 million people)<br>• Flood protection for ten million people<br>• Improved navigation above lake | • Lake and ports will silt up (530 million tonnes per year)<br>• 1.2 million people moved<br>• Loss of land<br>• Increasing deforestation, so erosion is worse<br>• Cranes and white flag dolphin threatened<br>• Scenic beauty and heritage lost |

**Case study 11    THE VOLTA RIVER**

The San Gabriel River (*Case study 6*) showed how changes in rivers have consequences for coasts. The damming of the Volta River in Ghana is another example.

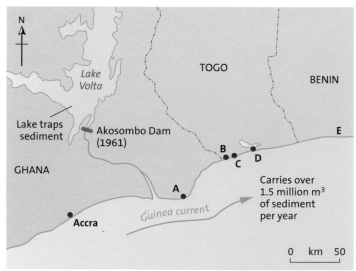

A The town of **Keta** is falling into the sea because the beach is being eroded

B The breakwater at **Lomé** is preventing sand from travelling eastwards

C The holiday resort of **Tropicana** is losing its beach and its tourist income

D The jetty at **Kpeme** is being undermined (this port exports phosphates)

E The coastal oil wells of **Benin** are now being threatened

**Figure 47**
*The problems of erosion east of the Volta estuary*

| Mega-dam | Year of completion | Advantages | Disadvantages |
|---|---|---|---|
| Akosombo (Volta) | 1961 | • Prestige project<br>• Cheap electricity<br>• Skilled employment<br>• Industrial growth (aluminium)<br>• Fishing (better diet)<br>• Improved water supply<br>• Lake transport<br>• New exports | • Loss of land<br>• 80 000 people moved<br>• Swamp land created<br>• Main road now a ferry<br>• Few top jobs available<br>• Profits go overseas<br>• Smelter uses most of the power<br>• Aluminium exported, so not used in Ghana<br>• Dam prevents sediment creating coastal beaches<br>• Increased coastal erosion<br>• Increased costs for resorts and ports |

**Table 5**
*Advantages and disadvantages of the Akosombo Dam*

## 22 Question

**Using case studies**

(a) Suggest how river and coastal management have contributed to this coastal erosion problem.

(b) Which of the following schemes do you think is
  (i) most beneficial?
  (ii) most harmful?
  Justify your decision.
  ■ Mekong dams
  ■ Aswan Dam (Nile)
  ■ Three Gorges Dam (Yangtze)
  ■ Akosombo Dam (Volta)

## Guidance

(a) The map key identifies the coastal problems that have resulted from the dam interrupting the supply of sediment to the coast. With little sediment available, erosion by longshore drift has increased massively.

(b) Try to justify your decision in terms of costs and benefits, the environmental impacts and the types of country involved.

# Modern river management

In recent years, we have begun to realise that **hard engineering** methods are too expensive, often used inappropriately and damaging to the natural environment. What is needed is an approach that works with natural processes and protects the environment.

**Soft engineering** schemes retain habitats and promote river ecosystems, allowing some flooding to take place. Some controversial projects attempt to 'turn back the clock' and undertake **river restoration (rehabilitation)**. **Wetland management** is a further step along the road to sustainable river management.

The effects of global warming and climate change mean that flood frequency and storm intensity will increase the risks to people and property. The increased roles of the Department for Environment, Food and Rural Affairs (DEFRA) and the Environment Agency, together with greater public awareness and tougher attitudes towards pollution, are also changing views. There is a growing debate about how best to tackle these issues and a move towards more realistic solutions — choosing the best strategy for each situation. **Integrated catchment management** is one such approach.

*Figure 48*
*Stone groynes in the Lune gorge near Halton — an example of hard engineering*

# Hard engineering

In the past, most river management strategies have involved considerable financial and environmental costs. These large schemes had unforeseen downstream impacts. Set against these concerns were the undoubted benefits of flood protection and economic advantage. A range of these structural solutions is summarised in Table 6.

*Table 6*
*Hard engineering schemes*

| Strategy | Benefits | Costs |
|---|---|---|
| Dam construction | • Regulates flow<br>• Stores water supplies<br>• Allows irrigation<br>• Generates electricity<br>• Prevents flooding<br>• Allows recreation | Upstream:<br>• Floods land upstream<br>• May displace population or habitats<br><br>Downstream:<br>• Less sediment and water flow<br>• Clearwater erosion<br>• Brings environmental change |
| Realignment/diversion | • Provides alternative<br>• Straighter and shorter route<br>• Steepens gradient<br>• Aids navigation<br>• Reduces flood risk<br>• Allows some control<br>• Habitats may survive | • May increase downstream velocity and erosion<br>• Flood risk may not be improved overall<br>• Interrupts riffle and pool sequence<br>• Habitat damage |
| Channelisation | • Increases river velocity<br>• Efficient<br>• Reduces friction<br>• Prevents sediment storage<br>• Halts migration and erosion<br>• Protects upstream from flooding | • Totally artificial environment<br>• Destruction of most river habitats<br>• May increase flood risk downstream |
| Revetment | • Largely artificial (various materials can be used)<br>• Protects banks from erosion<br>• Controls migration<br>• Limits land loss (may include levée-building) | • Damages bank habitats<br>• Prevents floodplain development<br>• Changes cross-profile |
| Wing dykes (groynes) | • Deflect thalweg, so keep channel clear<br>• Foster deposition below dykes<br>• Allow habitat growth<br>• Mimic natural processes and landforms | • Reduce overall width<br>• Slow velocity<br>• May not help navigation in practice (Figure 48) |

# Soft engineering

Soft engineering offers an alternative approach to river management and its environmental credentials are strong. Various methods are available, some of which are sustainable on a local scale. Soft engineering includes land use management, wetland and river bank conservation and river restoration, which actually deconstructs previous hard engineering. In the forefront of this approach in the UK is the Environment Agency with its Catchment Management Plans.

Figure 49 provides a comparison between hard and soft engineering. Note that although there are differences in methods, ecology and land use, there are a few similarities too.

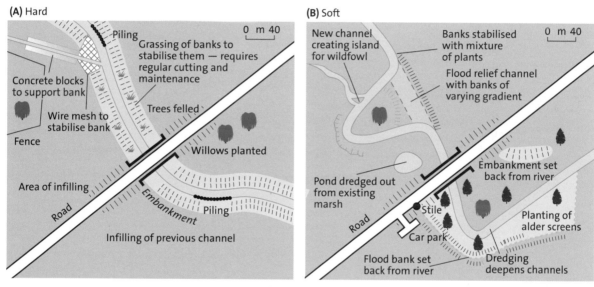

**Figure 49**
*A comparison of hard and soft approaches*

Examples of this softer approach, which copies or accommodates natural processes, are:

- **retaining natural features and ecosystems.** Following realignment or dredging, old meanders are often filled in, but this need not be so. They can be retained and planted with willows or poplars to give shade and anchor soil. Riffle and pool areas encourage biodiversity. Meanders can be fitted with a pipe inlet or sluice gate upstream and left open downstream to retain some flow (for example, on the River Witham in Lincolnshire). The banks of rivers not only contain the river but they have a large influence over the input of water from the land. River restoration exploits these features deliberately (examples include the Rivers Cole and Wharfe).
- **washland management** (flood storage ponds). These are areas of land deliberately allowed to flood seasonally. Often sited upstream of floodplain settlements, flood storage ponds help prevent flooding in 'at risk' locations downstream. Typically they are in flat areas supporting marshland or relatively poor pasture. There are examples near York, Shrewsbury and Nottingham. Figure 50 shows a storage area on the River Ribble as flood levels begin falling. Note the meanders, oxbow lake and natural levées built up higher than the floodplain.

**Figure 50**
Washland in Upper Ribblesdale, North Yorkshire

Bob Hordern

- **wetland conservation.** Wetlands, marshes and swamps are in serious decline globally, with half disappearing during the last century. Drainage for agriculture and water abstraction are the main reasons for these losses. Conservation could allow these areas to fulfil valuable functions, including:
  - flood control (see washland management above)
  - pollution control — absorbing 'dirty water' and filtering out pollutants
  - buffer zones — playing a key ecological role
  - habitat conservation — protecting important sites for wildlife, increasing biodiversity, and managing nature reserves and Sites of Special Scientific Interest (SSSIs)
  - streamway concept (see river corridors below)
  - low-input farming (such as grazing and reed production) and types of forestry and tourism recreation (sports, education and scenic value)
- **river corridors.** Land alongside the river is allowed to meander and even flood. This corridor concept complements flood storage and channel migration, allows recreation opportunities and supports green belt protection. The policy is to control planning and land use in river corridors in order to reduce the flood risk to people and property, while conserving the environment (see Figures 39 and 40).

## THE JUBILEE RIVER FLOOD RELIEF CHANNEL

*Case study* 12

Even in places where a structural response is needed, such as upstream of London, schemes can still be softened. To deal with the repeated flooding that has threatened 4800 homes, 700 businesses, major roads and the M4 motorway, an 11.8 km flood relief channel has been constructed. Leaving the Thames at Taplow Lock, it rejoins the river at Eton. Construction costs were £43.7 million, to be set against flooding costs of £52.7 million over 65 years. This is how long the scheme is likely to last before replacement.

The channel, now named the Jubilee River, is designed not just to alleviate flooding; it will also enhance the river environment and provide valuable local amenities.

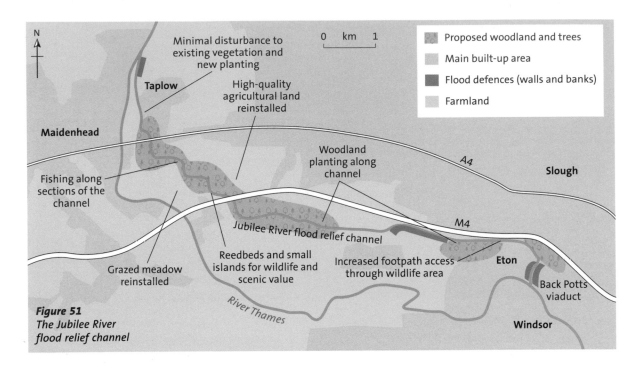

**Figure 51**
The Jubilee River
flood relief channel

## Question

(a) Using Figure 49, explain three ways in which scheme (B) provides more than flood protection.

(b) Using Figure 50, suggest how seasonal flooding may influence land use.

(c) Using Case study 12, calculate the cost-benefit ratio for the scheme.

## Guidance

(a) Environmental and recreational benefits are an added bonus with softer engineering schemes.

(b) Seasonal changes in washland areas reduce access to land as well as its value.

(c) Cost-benefit ratios are calculated by dividing benefits by costs.

# River and wetland restoration

River restoration means returning rivers to their former state. In the past, many streams were straightened to provide water power for mills, while larger rivers were engineered to allow floodplain settlement, more intensive agriculture or prevent flooding. Today's restoration schemes attempt to renew these damaged rivers and their habitats. Unlike hard engineering, these schemes are economical and small scale. Table 7 lists some examples, along with some useful websites to research.

In small streams, engineers use **in-channel devices** to restore the river environment and improve the quality of the water. Figure 52 shows the River Sowe near Coventry.

| Scheme | Website | Previous situation | Restoration schemes |
|---|---|---|---|
| River Kissimmee, Florida | South Florida Water Management District (www.sfwmd.gov) | • River channelised in the 1940s to reduce floods and aid navigation • Drainage of wetlands caused a 92% loss of birds, eutrophication from nutrients, salinisation of water supplies and reduced flow | • A major river restoration of 11000 hectares of wetland • This involves removing levées, infilling the canal and restoring meanders • Set to finish by 2012 • Progress is slow — hampered by the seasonal flow it is trying to encourage |
| River Danube, Eastern Europe | WWF (www.panda.org) | • Interventions resulted in 80% wetland loss and eutrophication levels 10 times those of the River Rhine • The Danube has 59 dams and is virtually a canal | • 'Green Corridor' plans include restoration of almost a million hectares of wetland along the floodplain • The Danube delta is also the location of a 46000-hectare Ramsar site (protected wetlands) • The crane population should recover |
| River Cole, Oxfordshire/ Wiltshire border | The River Restoration Centre (www.therrc.co.uk) | • Old leat (millstream) that often flooded • Slow-moving water, poor in quality • Low biodiversity | • Demonstration project involves reinstating old meanders and creating others along the old millstream • Improved habitat • Some seasonal flooding permitted • Reed beds and other features incorporated to help intercept silt and agricultural pollutants |
| River Skerne, Darlington | The River Restoration Centre (www.therrc.co.uk) | • Small river straightened to reduce flooding • Degraded by waste tipping to build a housing estate • Sewerage and power lines run alongside the stream | • Project involves reconstruction of old meanders, re-profiling streambed, and planting trees and wild flowers • Early results are encouraging and natural features are returning |

**Table 7**
*Four river and wetland restoration schemes*

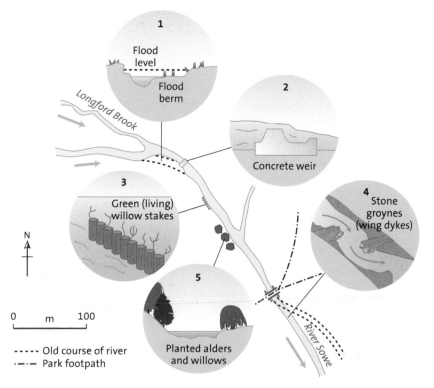

**Figure 52**
*In-channel devices, River Sowe, Coventry*

1 **Flood berm:** easily absorbs increases in discharge occurring at the confluence
2 **Concrete weir:** adjusts level of water, creates varied habitats, regulates flow and oxygenates water
3 **Green (living) willow stakes:** reduce flooding, prevent bank erosion and eventually grow (bio-engineering)
4 **Stone groynes (wing dykes):** encourage meandering, collect silt deposits, vary habitats and keep channel clear
5 **Planted alder and willows:** bind banks together, intercept surface and throughflow, shade water and add to biodiversity

The River Brede flows through the farmland of south Jutland in Denmark. In the 1950s, meanders were removed from the river to create a straight course to enable intensive livestock farming. Weirs in the river, as well as the straightening of its course, almost eliminated migratory sea trout.

An initial 5 km stretch of the river was re-meandered and, with the cooperation of farmers, a further 15 km length of the Brede has been restored as part of a nationwide strategy to improve the environmental management of river valleys. The Brede now meanders over its floodplain and seasonal flooding has been restored to the valley. This project obviously has important impacts for farmers and ecosystems in the floodplain.

In photo A, snow cover reveals the line of the now filled-in channel (marked T). The reconstructed meanders are marked R, and feature S is an oxbow lake created by engineers. Photo B, which was taken more recently, shows how the river looks today in the spring.

**Figure 53**
*The restoration of the River Brede*

The River Restoration Centre

**(A)** Winter        **(B)** Spring

---

**24** *Using case studies*

**Question**

(a) Visit one of the restoration websites in Table 7 and carry out further research into one of the four restoration schemes.

(b) Study Figure 53.
  (i) In the past, engineers created straightened sections along the River Brede, marked T on photo A. Suggest why this was done.
  (ii) Outline the negative consequences of straightening the river on the river environment.
  (iii) With the help of photo B, explain how river restoration has affected land use and ecosystems in the valley.

## RIVER RESTORATION IN THE YORKSHIRE DALES

This case study describes how river restoration can become a community activity and shows how a local flooding problem can be solved sustainably. It relates to the junction of Cray Beck and the infant river Wharfe at the head of upper Wharfedale near Buckden. Flooding results when streams suddenly enter this flat glaciated valley. This causes a meandering channel, deposition of river gravels and variable levels of discharge (Figure 54).

Pressure for action came from various stakeholders:

- local farmers — continual damage to land and boundary walls, and livestock losses
- Environment Agency and partners — need to manage hydrology and ecology, and be aware that gravel trap is defunct. This collects larger material in times of flood
- Buckden residents — much inconvenience and property damage
- National Park and partners — most able to access external funding

The following options were considered:

- hard engineering — channelisation to widen the channel, stabilise banks, speed up flow and keep river gravels moving
- do nothing — allow channel migration and flooding
- bypass channel — allow flood water and gravel to pass, which increases vegetation
- realignment — carry out some straightening, lower base levels, allow some flooding
- gravel management — carry out periodic gravel excavation, repair trap

All parties agreed on the proposal to work with the river, allowing some migration, strengthening small levées and planting trees on gravel deposits. This choice of scheme had many sustainable features but some engineering work was still involved.

Steep slopes
Flat floodplain
Stone revetment
Gravel bar regraded
Tree planting
Migration allowed
Pools and riffles retained
New fences to limit livestock access
Confluence with beck realigned

Bob Hordern

**Figure 54**
*Practical approach to the restoration of the River Wharfe near Buckden*

# Integrated catchment management

Although hard engineering is probably necessary to protect urban areas, there is an obvious place for sustainable management of floodplains in rural locations. Decisions about the most appropriate strategy require informed analysis. The Environment Agency has used its catchment research to put together and implement plans that consider the whole of a river's catchment. Ideally, these should be sustainable and attempt to establish:

■ a vision for the catchment (in a consultation report)
■ a policy to tackle issues that arise (outlined in catchment plans)
■ costed action plans to address the issues (Local Environmental Action Plans)

Using flood management as an example, different management responses are possible. Some responses may be designed to tackle the causes of floods, which may be due to weather conditions, the characteristics of the catchment and people's use of the land. Other responses may seek to modify the nature of these floods by managing the river itself.

It is important to raise public awareness of flood risks and to educate residents about how to be prepared. The Environment Agency's flood page at

**www.environment-agency.gov.uk/subjects/flood**

is a useful source of information.

River management strategies can be classed as:

■ **abatement** — including wider catchment strategies, afforestation, managing moorland, changing farming practices, building reservoirs and controlling urban runoff
■ **protection** — including structural operations such as building embankments, using diversion schemes, flood storage, barrages and flood barriers
■ **adjustment** — involving floodplain schemes such as emergency action, flood-proofing, land-use regulation (zoning) and insurance cover

## MANAGING THE YORKSHIRE OUSE CATCHMENT

- Together, the Ouse and Trent catchments drain one fifth of England into the River Humber.
- Yorkshire's rivers obtain most of their water from the Yorkshire Dales, where rainfall exceeds 1500 mm per year and snow cover can last up to 45 days.
- The drainage pattern focuses on the plain of York, which is a flat floodplain remarkably close to sea level and used for arable farming. York and a few smaller towns sit astride these rivers in the path of any flood.
- The great floods of 1947 and 1982 were particularly damaging for York; much has been done recently to combat risks and protect the city. In 1999, attention switched elsewhere in the catchment as the River Derwent burst its banks at Malton.
- The estuary area is a further worry; it is one of the areas of the UK at risk from inundation as global sea levels rise.

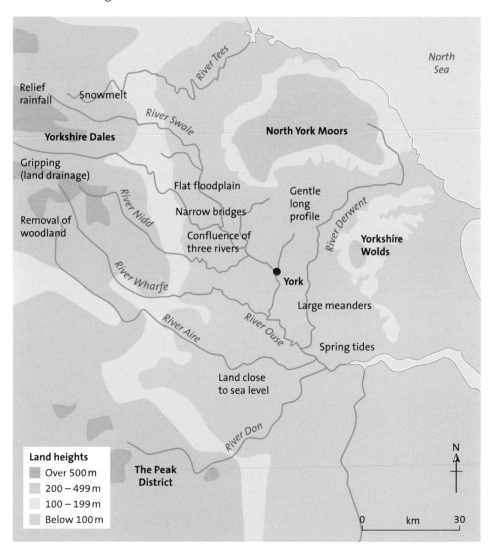

**Figure 55**
*The causes of flooding in the Ouse catchment*

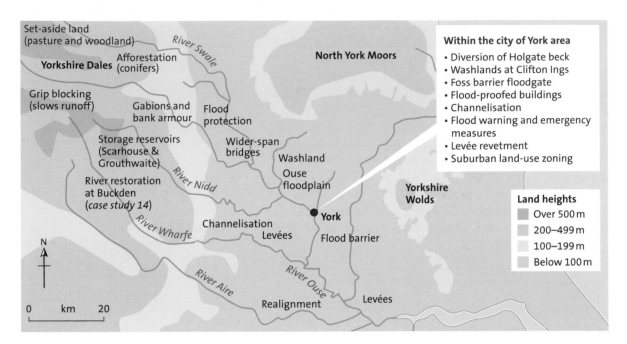

**Figure 56**
*The management response to flooding in the Ouse catchment*

Integrated catchment management involves a wide variety of issues. In the upper courses of the Trent and Ouse, managers are concerned with hydrology and water storage, while the more complex matters of flood control, abstraction and pollution dominate decisions about their lower courses. Figures 55 and 56 show the causes of flooding and the already wide range of methods employed to manage flooding in this large river catchment.

## Question

(a) Using Figure 55, divide the causes of flooding into those that are due to:
    (i) weather
    (ii) river catchment features
    (iii) people's use of land

(b) Carry out a similar exercise for flood management using Figure 56 and the following categories:
    (i) abatement
    (ii) protection
    (iii) adjustment

## Guidance

Arrange the headings into columns and refer to Figure 38 to help you understand the causes of flooding.

# Coastal systems

Coasts are subjected to marine processes, especially from waves, as well as the terrestrial processes discussed in Section A. Coastlines are dynamic and relatively complex environments, not only because they undergo change in the short and long term, but because — as attractive environments for development — they are under increasing human pressure. Coasts can be seen as a zone where a number of geomorphological systems interact. In system terms, rivers input fresh water and sediment, waves and currents provide energy, and cliffs and beaches respond to these changes.

Changes to coasts take place within a range of timescales. The effects of tides, a single storm or a management decision reveal themselves relatively quickly. Many other changes take longer, slowly shaping landforms along the coast. Some are linked to the ice age or the period following it. They are also affected by longer-running processes relating to changes in sea level.

It is important to identify the range of factors that affect coasts (Figure 57).

**Weather and climate**
• Wind strength and direction
• Rainfall and temperature
• Storms and surges

**Human activities**
• Intervention into natural systems
• Use of land for development

**The nature of coasts**

**Land**
• Shape of the coastline
• Relief of the land
• Presence or lack of a beach
• Structure of the coast
• Resistance of the rocks
• Sub-aerial processes
• River sediments
• Coastal (land) ecosystems

**Sea**
• Coastal (marine) ecosystems
• Wave energy and direction
• Size and type of waves
• Local currents and longshore drift
• Tidal changes
• Water depth
• Offshore sediments
• Longer-term changes in sea level

**Figure 57**
*Factors influencing coastal change*

The influence of rivers and sub-aerial factors is discussed in Section A; human impacts will be examined later.

# Waves, tides and currents

The marine offshore system is driven by the effects of waves, tides and currents.

## Waves

Waves are the main source of energy along coasts, being created either when winds blow for great distances over the surface of the sea (the drag effect creates a swell) or when local weather conditions produce storm waves. Exposed coastlines, such as those in the southwest of England, experience waves whose **fetch** began several thousands of kilometres away.

There are many terms relating to wave motion and characteristics including wavelength (the distance the waves are apart), **wave frequency** (how many waves arrive per minute) and wave height. Wave steepness may well be an important factor in determining the effects waves have. When waves reach the coast, they 'break' and water rushes forward as **swash** before returning as **backwash**. This happens because of the drag effect of the shallowing water. Water returning down the beach is often concentrated into **rip currents**.

Wave refraction occurs when the configuration of the coast interferes with the pattern of waves. This helps to explain why bay beaches attract deposition while deeper water around headlands favours erosion. A lot of theory and research exists about waves — studies are carried out both outdoors and in the laboratory.

Classifying waves as constructive and destructive types allows us to gain a basic understanding of these complex phenomena (Figure 58).

**(A)** Constructive

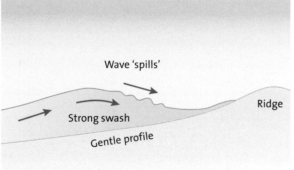

**(B)** Destructive

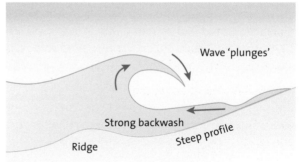

*Figure 58
Constructive and
destructive wave
types*

■ **Constructive waves** are flat (they have a long wavelength and low height) and their strong swash carries sand up the beach. This creates a shelving profile that allows water time to percolate rather than run off. These 'spilling' or 'surging' waves create a ridge or berm at the top of the beach.

■ **Destructive waves** are steeper and often occur in high-energy locations or during storms. Because of their height, they 'plunge' at the beach, removing sediment in their backwash. Ridges tend to form at the seaward end of the beach.

## Tides

Tides are important as they set the level at which the sea operates. Tidal ranges vary considerably around the UK, with Kent and Sussex reaching over 4 m, while Dorset and west Wales have relatively low figures. Spring tides are especially worrying in low-lying estuaries as they can lead to coastal flooding, especially when combined with storm surges. Strong tidal currents can also be a problem for shipping. Areas where the tidal range is small (less than 3 m) tend to contain depositional features like spits, bars and sand dunes.

## Currents

Although there are global currents that respond to the world pattern of ocean temperatures, prevailing winds are most likely to be responsible for water movements along coasts. Winds drive waves shorewards at an angle to the shoreline. Along the south coast of England, waves are typically from the southwest, whereas along the Yorkshire coast, they arrive from the northeast. The most common type of current is therefore **longshore drift**, which is examined more closely in Part 7.

# Sediment systems

For research and management purposes, the coastline of England and Wales has been divided into a series of 11 **sediment cells** (Figure 102) where the inputs and outputs appear largely balanced. If these units are broken down into smaller sub-cells, details of local movements can be more closely studied and managed. In these sub-cells, sediments derived from rivers, cliff erosion or offshore sources are transferred along the coast. This material may be deposited in sinks such as beach ridges, sand dunes and spits, or carried away to offshore banks.

To give an idea of scale, longshore currents within the Holderness coast sub-cell in Yorkshire transport over 500 000 tonnes of sediment southwards annually, mostly in suspension (Figure 59).

The foreshore zone is occupied by beaches, dunes, mudflats and coastal ecosystems. The systems approach can also be applied here. Beaches can be seen as stores, continually importing and exporting material. For a beach to remain stable, there needs to be a balance of processes (Figure 60). However, rising global sea level and increased human interference make this state of dynamic equilibrium increasing unlikely. It is perhaps inevitable that some of these human processes will be significant locally.

*Figure 59*
*The sediment system on the Holderness coast*

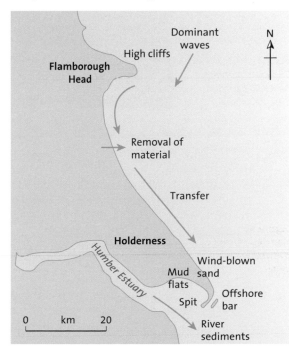

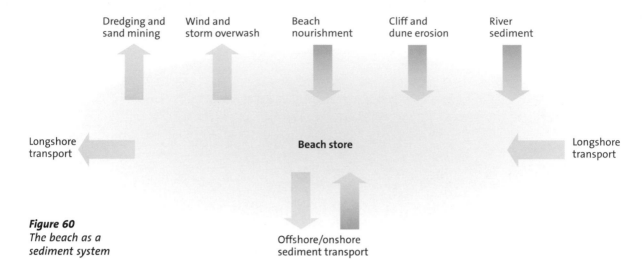

**Figure 60**
*The beach as a sediment system*

**Figure 61**
*The coastal 'battle-ground', Hornsea, East Yorkshire*

Bob Hordern

**Using case studies**

## Question

Study Figure 61.
(a) Identify the factors at work on this coastline.
(b) Which factors appear to be winning this battle?

## Guidance

(a) Use Figure 57 to help you to select the factors that are relevant to the photograph.
(b) An example would be storm winds (driving waves on to the beach — waves then damage defences).

# Coastal landforms and processes

Coastal landforms have been characterised traditionally as being about 'capes and bays', a view that dates from a time when geography was largely about scenery and maps. However, there are many other coastal landforms and ecosystems that deserve attention.

## Coastal erosion

Erosional processes along coasts are of a broadly similar nature to those seen in river environments in that they involve water and sub-aerial processes (see Part 2), although there are some important differences.

■ **Terrestrial processes** are best seen as a series of sub-systems, including weathering, mass-movement and surface runoff. The effects of these are most noticeable along cliffs with massive joints (limestone) or where water encourages gullies and landslides (clay).

■ **Marine processes** rely on the nature of the waves acting at the foot of the cliffs. Again, the geology of the cliffs may accelerate or delay the rate of erosion. Hydraulic action or wave quarrying is the main erosive force, with wave impacts reaching a maximum in storm conditions. Additionally, where air is trapped in joints and fissures, the resulting pneumatic pressure weakens cliffs, causing the collapse and removal of large blocks. **Abrasion** (corrasion) results from waves throwing material against cliffs and is most effective when large pebbles and high-energy waves are involved. Attrition progressively rounds off and reduces the size of pebbles and sand grains as they are moved back and forth continually by waves.

Headlands and bays tend to form where there are alternating resistant and weaker rocks or where there are strong patterns of faults or joints. The distinctive structure of the Purbeck coast of Dorset has a sequence of limestone and clay rocks. These produce discordant features on the eastern foreland (where the rock structure is at right angles to the coast) and concordant (parallel) features on the south-facing coast around Lulworth Cove. Once established, headlands become the focus of intense local erosion as relatively deeper water and the refraction of waves exploit any weaknesses in lithology.

Make sure you use the following words carefully, as they have slightly different meanings:

- **Structure** refers to the arrangement of rocks in the landscape and includes features such as strata (layers), dip (angle of rocks) and faults.
- **Lithology** is about the composition of the rock itself, which determines how vulnerable it may be to chemical or mechanical damage. A rock's permeability is often controlled by its grain size and the joint patterns or bedding planes within the rock.

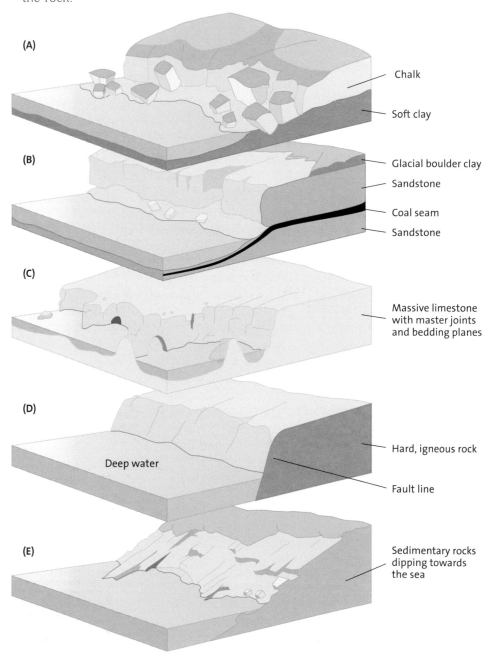

(A)

Chalk

Soft clay

(B)

Glacial boulder clay

Sandstone

Coal seam

Sandstone

(C)

Massive limestone with master joints and bedding planes

(D)

Deep water

Hard, igneous rock

Fault line

(E)

Sedimentary rocks dipping towards the sea

***Figure 62***
*A range of cliff profiles*

## 28 Question

Study Figure 62.
(a) Explain why cliffs in profile A are likely to erode easily.
(b) Explain how the lithology of profile B is producing the cliff profile shown.
(c) List the coastal landforms evolving in profile C (see also Figure 64).
(d) Suggest reasons why cliff profile D might erode least rapidly.
(e) Suggest why the coast in profile E might erode most rapidly.

### Guidance

These questions emphasise the links between rocks and structure and how cliffs are eroded. Focus on the labelling and the shape of the cliffs.

## THE CLIFFS AND HEADLANDS OF ORKNEY

*Case study* **16**

Orkney is located off the northeast coast of Scotland and has a variety of coastal features. At its highest point, it is only 477 m above sea level and consists of a large, central island — the Mainland — surrounded by perhaps as many as 70 other smaller islands. Figure 63 shows how earth movements and geological faults have helped break up Orkney and created an inner area of water called Scapa Flow.

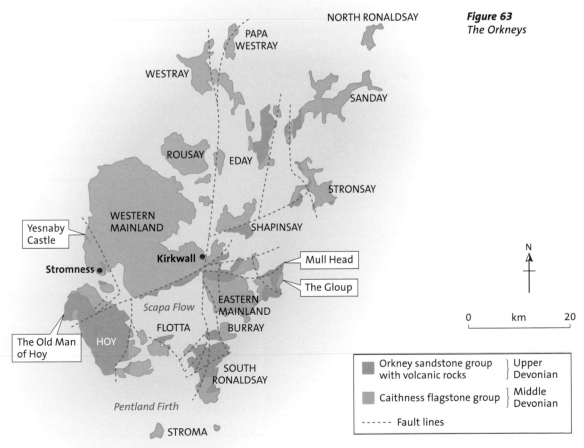

**Figure 63**
*The Orkneys*

**Figure 64**
*Factors leading to micro-features in massive resistant rocks*

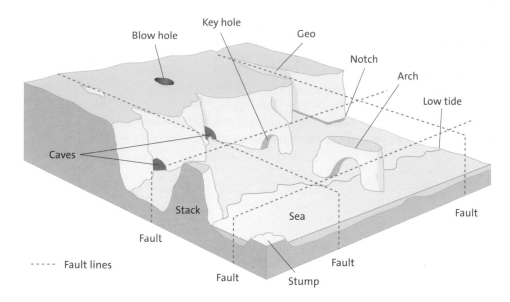

Most of Orkney is made up of sedimentary sandstones that produce distinctive layers in many coastal cliffs. Elsewhere there are large areas of thin-layered flagstones that have been more easily eroded, especially where there are local faults and folds. Some volcanic lava flows are found in the rocks of Hoy. Above these, soft boulder clay and glacial erratics were later deposited during the ice age.

The coastline of Orkney is over 800 km long and includes steep cliffs, sloping wave-cut platforms and small beaches. The continual effects of waves, wind and winter storms along exposed coastlines have led to rapid erosional changes. Much of the Atlantic (west) coast is characterised by high cliffs. A series of headlands from Stromness northwards shows how the interaction of processes and geology produce a cycle of cliff features from the erosion of initial weaknesses to the formation of a whole series of micro-features, as shown in Figure 64.

Three particular locations in Orkney are worth more detailed study:

■ The Old Man of Hoy is perhaps Orkney's most famous landmark.

■ Yesnaby Castle on the western Mainland is another striking example from this coastal erosion sequence.

■ Mull Head on the east coast has many similar, though more rounded, features, including the intriguing Gloup blowhole.

## The Old Man of Hoy

The Old Man of Hoy (Figure 65) is Britain's tallest sea stack. It is 137 m high but only 30 m wide at

**Figure 65**
*The Old Man of Hoy*

www.undiscoveredscotland.co.uk

its base, and its sheer sides have been eroded from Orkney's sandstone rocks. The changing shape of the stack has been well documented in paintings and maps. In 1750, the stack was depicted as a headland. By 1820, erosion by westerly winds and wave action had carved an arch — the two legs giving the Old Man his name. Today, only the single outer pillar of rock remains.

Rock ledges in the stack result from differential weathering of the soft sandstones and more resistant flagstones. Joints and faults divide these sedimentary rocks into blocks, which have helped retain the rectangular shape of the feature. The stack is some 60m away from the main coastal cliffs and the intervening area is covered in rock debris. The larger material is from the collapsed arch, while the steeply sloping screes result from recent weathering and cliff rockfalls, accelerated by the high local rainfall and freezing temperatures of winter. The stack sits on a plinth of dark basalt lava and older rocks, helping to preserve the surrounding wave-cut platform.

Rapid changes threaten to destroy this feature. Weathering of the upper section is already exploiting a 40 m vertical crack in the rock, while waves continue to erode the base. Here and throughout the west coast there are few beaches to provide protection. The powerful swell of Atlantic waves (long wavelength and fetch), the relatively deep water offshore, and the steep wave-cut platforms create a high-energy coastline where erosion dominates.

***Figure 66***
*Yesnaby Castle*

Sue Warn

## Yesnaby Castle

Yesnaby Castle is a distinctive 35 metre-high sandstone sea stack north of Stromness. Above the base there are two pillars of rock divided by an arch, cut by wave action in storm conditions. Like the Old Man of Hoy, it is a favourite spot for climbers and fossil collectors.

## 29   Question

*Using case studies*

(a) Using a copy of Figure 65 or a rough sketch and the account of the Old Man of Hoy, label the features shown on the photograph.
(b) Using a different colour, write annotations to explain the main processes involved.
(c) Explain where Yesnaby Castle (Figure 66) fits into the 'cycle of cliff erosion'.

### Guidance

Annotations are an effective way of exploring how landforms are made, and can be a useful revision aid. Refer to Figure 64 when answering these questions.

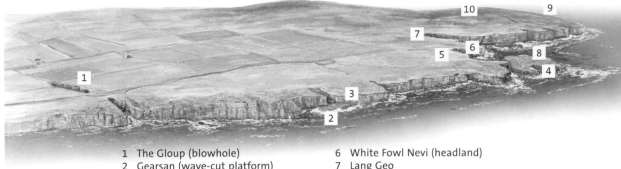

1  The Gloup (blowhole)
2  Gearsan (wave-cut platform)
3  Clu Ber (headland)
4  Brough of Deerness (future island)
5  Large Burrah Geo
6  White Fowl Nevi (headland)
7  Lang Geo
8  Howan Lickan (wave-cut platform)
9  Mull Head
10 Covenanter Memorial (gunnels in cliffs)

**Figure 67**
*Mull Head nature reserve*

## Mull Head

The Deerness peninsular forms the northeastern corner of the mainland of Orkney. At Mull Head, many of the micro-features of Orkney's coastline are seen in a nature reserve surrounded by 160 hectares of sea cliffs (Figure 67).

■ The cliffs are made up of two different types of flagstone — Rousay and Eday. Thin rock layers enable the weather and sea to exploit weaknesses while seabirds nest on the resulting horizontal ledges.

■ The exposed northern cliffs have been worn into smooth gunnels by the driving winds and rain from the Atlantic Ocean.

■ On the southern side, faults have been opened up by wave quarrying to form caves and geos (for example, Lang Geo), which run inland between intervening narrow headlands and small wave-cut platforms.

**Figure 68**
*The Gloup looking seawards*

www.undiscoveredscotland.co.uk

■ The Brough of Deerness is a minor part of the headland about to be detached to form an island. Waves are refracted around its sides, focusing their power on the remaining narrow landward neck.

■ Minor faults in the rocks are the main reason for local weakness, seen spectacularly at the Gloup blowhole (Figure 68). The name 'Gloup' comes from the Gaelic *gluppa*, meaning 'chasm'. Its formation occurred as follows:
  – Initially, a narrow cave developed along a vertical fault in the cliff face.
  – Erosion by water took place as a result of hydraulic action and air was forced inland by incoming waves.
  – Increased damage to the cave roof lead to repeated rockfalls and, eventually, a collapse, which allowed waves to force water upwards, so creating a blowhole.
  – Further erosion and rockfalls opened up the landward side, leaving a 25 m-deep geo.
  – Part of the roof remains as a rock arch connecting the Gloup with the open sea.

## 30 Question

Using located examples, explain the effects of rock structure and lithology on cliff coastlines.

### Guidance

This is a classic examination essay question. Prepare a plan, using information from *Case study 16*. Refer to features, choosing from:

(a) the Old Man of Hoy
(b) Yesnaby Castle
(c) Mull Head
(d) the Gloup

# Longshore processes

Many features of coastlines are formed by longshore drift, which transports sand and shingle along the coast.

If waves strike the coast at an angle, the returning water follows the steepest path (i.e. at right angles to the coast). Repetition of this results in a zig-zag pattern of water and sediment movement. This produces a net migration of beach material downdrift (along the beach) — see (Figure 69). An angle of 30° to the shoreline seems to cause maximum erosion.

Beaches affected by this process are said to be drift-aligned. Though fairly constant, this process occurs more rapidly in storm conditions. **Groynes**, which interrupt this process by catching sediment, provide clear evidence of its presence.

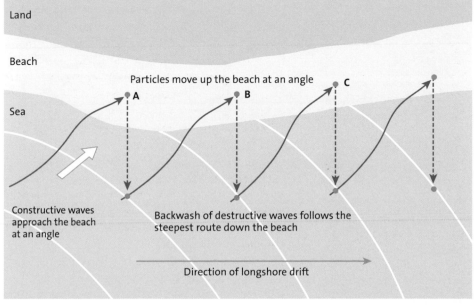

*Figure 69*
*Longshore drift*

Land

Beach

Sea

Particles move up the beach at an angle

A

B

C

Constructive waves approach the beach at an angle

Backwash of destructive waves follows the steepest route down the beach

Direction of longshore drift

**A** First position of pebble  **B** Second position of pebble  **C** Third position of pebble

## LONGSHORE DRIFT IN PORLOCK BAY

**Figure 70**
*Porlock Bay, looking west from Hurlstone Point to Gore Point*

Porlock Bay is one of the last features of the Somerset coastline before reaching north Devon. A shingle beach stretches 5 km across the bay between the headlands of Gore Point and Hurlstone Point, protecting the marshland behind from being flooded by the sea.

Longshore drift carries material west to east across the bay (towards the camera in Figure 70), sorting the pebbles and creating different beach profiles at either end of the beach. Pebbles at the eastern end (foreground) are generally rounder, smaller and less varied in size than those found in the west.

Groynes were built to prevent long-shore erosion, but this led to changes further east. Today, the weakening of the shingle ridge, the effects of rising sea level and severe storms all threaten the marsh and farmland behind.

Mark Bolland, FSC, Nettlecombe

### Fieldwork data

Fieldwork data about Porlock's shingle beach have been collected over many years. Data from the Field Studies Council are shown in Figures 71 and 72.

**Figure 71**
*Pebble calibre and shape*

**(A)** Pebbles at Gore Point (west)

**(B)** Pebbles at Hurlstone Point (east)

Mark Bolland, FSC, Nettlecombe

**Figure 72**
*Beach profiles*

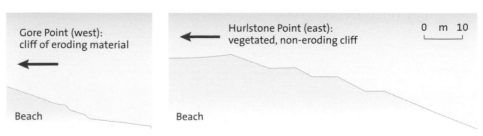

Gore Point (west):
cliff of eroding material

Hurlstone Point (east):
vegetated, non-eroding cliff

0   m   10

Beach

Beach

## Question

Using the data from Figures 71 and 72, copy and complete the following table.

| Figure reference | Gore Point (west) | Hurlstone Point (east) |
|---|---|---|
| Figure 71 Pebble calibre and shape | The pebbles in the west at Gore Point are... | The pebbles in the east at Hurlstone Point are... |
| Figure 72 Beach profiles | At the western end of the bay, Gore Point's beach... | The beach at Hurlstone Point is more substantial. Deposition, shown by the well developed backslope, has created a steeper shingle ridge, now colonised by vegetation. The beach ridge is over twice the height of that at Gore Point and extends landwards some 70 m. |

## Guidance

In the first row, describe the size and shape of pebbles at the two sites. In the second row, complete the sentence by using the answer for Hurlstone Point as a guide.

## Secondary data

Secondary sources, such as old maps, photos and written documents, can provide valuable information. Local history groups and university departments often have dedicated internet sites. Historical information about the construction of coastal defences is often well documented and reports by coastal engineers may also be available.

Evidence at Porlock reveals that human interference by various management schemes may have had unexpected impacts:

■ A major groyne built to protect the harbour at Porlock Weir has led to **beach starvation** further east, and other groynes built in the bay in the 1960s now appear to be failing due to wave scour. This engineering has changed the direction of wave attack (refraction) and caused a thinning of the beach.

■ The New Works Outfall, designed to control the size of lagoons in the marshes by allowing freshwater to escape through a drain into the sea, has created a 'bulge' in the beach. The western beach is now even more fragile.

■ During a storm in October 1996, the ridge was breached and seawater flooded the marshes. This will keep happening as sea level rises.

Longshore drift has led to a coastal flooding problem caused by poor management. There is now a growing threat to farmland, a valuable habitat and the area's tourism potential. The question is how best to manage this problem.

*Figure 73*
*A student's sketch map of Porlock Bay*

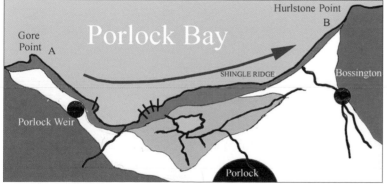

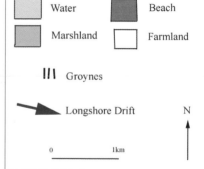

**Question**

(a) Explain how natural processes have created the landforms and pebble features of Porlock Bay.

(b) What does the secondary data tell you about the effect of people interfering with these natural processes?

(c) Carry out further research into the area using the Porlock Tourist Association website (**www.porlock.co.uk**), or **www.stacey.peak-media.co.uk/PorlockBay/PorlockBayAug2005.htm**.

**Guidance**

Parts (a) and (b) should provide you with a sound case study to use in examinations, while part (c) will provide you with a better understanding of the area.

# Coastal deposition

It is not surprising that with 70% of coasts worldwide suffering net erosion, and a continued global rise in sea level, only 20% of the world's coastlines are depositional. Deposition occurs in low-energy environments where the effects of waves, storms and tides are reduced. A range of coastal features result from **coastal deposition** — beaches are perhaps the most obvious. These belong to two broad categories:

■ **Swash-aligned beaches** are formed when waves break parallel to the coast. These produce bay beaches (such as Lulworth Cove in Dorset) and barrier beaches (such as Start Bay in Devon).

■ **Drift-aligned beaches** are formed when longshore drift moves material along the coast, producing elongated and partly detached features. Spits like the one at Spurn Head in Yorkshire (*Case study 18*) are created in this way.

Recent evidence from studies of beaches such as Chesil Beach in Dorset and Start Bay in Devon suggests depositional processes may be more complex than once thought. Modern research indicates that many of our shingle beaches were largely constructed following the ice age. Sea levels were much lower during glaciation. Afterwards, when the ice melted and sea levels rose, the waves (swash) brought loose material onshore, creating beaches. Over time, this source of supply has reduced and so now longshore drift is the dominant process, largely re-sorting earlier beach deposits.

## Beach geomorphology

There is a direct relationship between wave power and the calibre of beach material. Low-energy locations are likely to receive more sediment, and this is usually of a smaller calibre. The same is true of beach gradient — sand produces shelving beaches, whereas shingle beaches may approach angles of 30°. Larger material collects in the upper part of the beach because returning backwash deposits heavier particles first, while finer material is carried seawards. Beach drift moving along the shoreline may also produce a progressive fining of material, as seen at Porlock Bay (*Case study 17*). Closer examination reveals surface features called beach landforms (Figure 74).

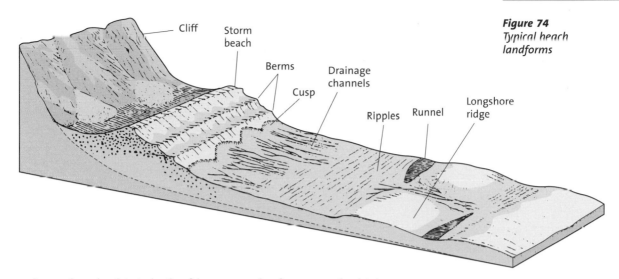

Figure 74
*Typical beach landforms*

- Storm beach: this is built of large, angular fragments by high-energy waves.
- Berms: these ridges are formed at high tide level by constructive waves.
- Cusps: these small embayments are not easily explained and are found at the top of the beach where the gradient begins to steepen.
- Drainage channels and ripples: these are small features formed in sand and linked to mid-beach drainage. Channels may cut through the longshore ridge.
- Runnels: in the lower part of the beach these reveal how the water has progressively spread out across the beach as the tide has fallen.
- Longshore ridge: this is the most seaward ridge, where destructive waves have dragged material from the beach.

Although many beaches seem to follow the model in Figure 74, coasts are not made up of only beaches but include a range of shoreline features.

Figure 75
*Shoreline features at Robin Hood's Bay*

Bob Hordern

**33** **Question**

Study Figure 75.
**(a) Identify four shoreline features shown.**
**(b) Explain the distribution of the various beach deposits.**

**Guidance**

**(a)** These features occur in four bands across the photograph.
**(b)** Beach deposits depend on the effects of tides, weather conditions, geology and the size of the material involved.

*Using case studies*

*Rivers & Coasts*

A range of drift-aligned or detached beaches can be found around the UK. These occur in low-energy environments where waves rather than tides prevail. Landforms such as spits tend to occur at the end of a coastal sub-cell, where there is a sudden change in coastal configuration. Other features — such as bars, tombolos and cuspate forelands — may be similarly formed (see Figure 76). Table 8 lists some examples of these classic landforms.

**Figure 76**
*A range of depositional beach landforms*

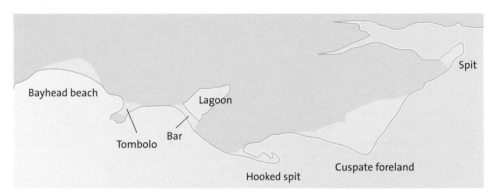

**Table 8**
*Depositional landforms*

| Landform | Example | Context/process |
|---|---|---|
| Bayhead beach | Lulworth Cove, Dorset | An enclosed bay with waves arriving at right angles to the shore. Deposition increases in quiet water while erosion is refracted laterally into the headlands. |
| Tombolo | Great Orme, Llandudno | A spit-type growth of sediment that may attach an island to the coast. Alternatively, the shelter of the island may encourage deposition. |
| Lagoon | The Fleet, Chesil Beach, Dorset | These form behind bars. They may be saltwater or freshwater and are likely to become silted up and colonised by successive vegetation. |
| Bar | Looe, Cornwall | Bayhead bars may be where a spit has grown across a bay, connecting it to the opposite shore. Alternatively, they may be swash-aligned. |
| Hooked spit | Spurn Head, Yorkshire | As spits build out into deep water, they require increasing volumes of sediment to build above the high mark. The tip or distal turns towards the land where it is shallower. Once formed, hooks are sheltered from the dominant waves by the spit. |
| Cuspate foreland | Dungeness, Kent | Changes in the pattern of growth of spits may cause larger, more complex features to form, especially where tidal effects are weaker. |
| Spit | Orford Ness, Suffolk | A ridge of sand or shingle attached to land at one end and finishing in open sea at the other. Spits migrate and extend but they may also become detached in storms. |
| Barrier beach | East Coast, USA | An extensive, bar-type landform probably created by post-glacial swash, now migrating inland from a hurricane-prone coastline. Less common in Europe, but a feature of the Netherlands and Baltic coastlines. |

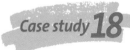 *Case study 18* — SPURN HEAD

Spurn Head is a classic spit landform located at the entrance to the Humber estuary. It has formed from glacial material eroded from the Holderness cliffs and carried southwards by longshore drift, a process that has continued since the end of the ice age.

This landform is a long, narrow extension of sand and shingle projecting out into the estuary. It is this change in the direction of the coast and the arrival of river discharge that triggers the deposition process. The distal end of the spit curves southwestwards because of wave refraction and tidal currents.

Spurn is currently about 6 km long, migrating westwards and growing in length by about 10 cm per year. Spits are dynamic and prone to change (see Figure 77). In the case of Spurn, history suggests that the feature has a growth cycle of about 250 years. After this period, the sea, driven by northeasterly winds, will break through at the narrow neck and cause the spit to become temporarily detached. This happened in the fourteenth, seventeenth and nineteenth centuries but, ominously, there were a number of small scale breaches throughout the 1990s.

In the sheltered Humber estuary, tidal deposition has created mudflats with a narrow fringe of saltmarsh. Sand blown from the shrinking shingle beaches on the exposed North Sea side of the spit has been built into a single line of dunes, colonised by marram grass.

Further research into this location can be carried out using the link to the 'Spurn Point Heritage Coastal Map', from the East Yorkshire Coastal Observatory website at **www.hull.ac.uk/coastalobs/ resources.html heritage** and its links to the picture gallery at **www.panoramic-imaging.com/eastriding/ spurn.htm**.

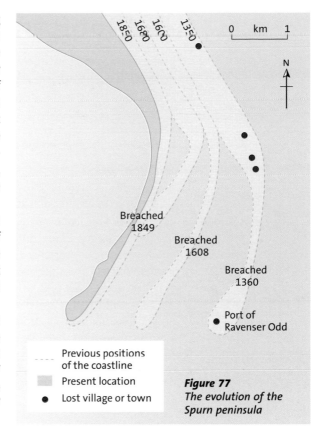

**Figure 77**
**The evolution of the Spurn peninsula**

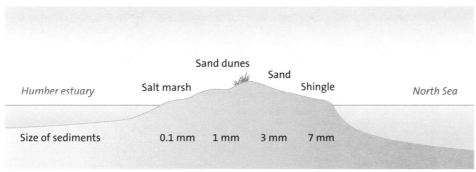

**Figure 78**
**A cross-section of Spurn Head**

## 34 Question

Write a summary of how the Spurn Head has formed, referring to:
(a) location
(b) landform shape
(c) sediment sizes
(d) ecosystems

### Guidance

This tests your understanding of processes and the idea that landforms are continually changing.

# Changing sea levels

Waves, tides and storms affect coastlines in the short term, but there are longer-term changes that can have greater impacts. It is best to look first at the processes behind these changes and then at the landforms they create.

## Processes

The processes that lead to long-term changes in sea level are global and local.

### Global changes

These **eustatic** changes relate to temperature. Global warming and the ice age are part of this phenomenon, which involves changes in the balance between the oceans and the ice caps.

- 18 000 years ago, when temperatures were about 5°C below those of today, ice still covered much of the Earth's surface and so there was less water in the oceans. Sea level was perhaps 150 metres below its level today. After the ice age, global temperatures and sea level rose steadily, creating today's coastline. Over the last century, sea level rose by about 15 cm, and in recent years there is growing evidence of an increase in global warming. Low-lying coastal locations could face an increased risk of submergence: a sea level rise of a further 50 cm throughout the twenty-first century is an often-quoted estimate.
- Rising sea level means less land and coastline retreat. The extent of this retreat can be seen by looking at a physical map of the world in an atlas, where it is seen as a light-coloured shelf of shallower water around the world's coasts. As sea level rose and water flooded the lowlands and valleys, areas of higher land became isolated and formed islands. This helps explain what happened to the British Isles as they became submerged, separating Great Britain first from Ireland and then from mainland Europe as the North Sea flooded.

### Local and regional changes

In some parts of the world, land may move up or down relative to the sea, and this will have an effect on the coastline. The causes of these changes vary but are often referred to as **isostatic** effects:

- During the ice age, parts of the Earth's crust were weighed down by the amount of ice covering the land. When the ice melted, the land emerged slowly again. This isostatic recovery continues today in many places, but it is most noticeable where the rate of recovery is faster than the global rise in sea level. This is seen in Canada and Scandinavia, where uplifts reach over 20 mm per year. In the British Isles, there appears to be a tilting effect, with the glaciated northwest rising and southeastern areas — which were not covered by ice — subsiding.
- The opposite effect is seen at the mouths of many of the world's major rivers as enormous quantities of sediment are building up in massive deltas. The colossal weight of this material presses down on the coastal and adjacent sea bed. Where the Mississippi River enters the Gulf of Mexico, the surface of the delta is slowly subsiding as the material within it is compressed.
- In other parts of the world, most obviously along the edges of the Earth's tectonic plates, there are other movements of the Earth's crust. Forces beneath the surface

may lift areas up relative to the sea. This occurs where the mountain ranges of North and South America meet the Pacific Rim. In contrast, some areas of the Mediterranean coast in Europe are being forced downwards.

## AMERICA IN PERIL FROM THE SEA

The title of this case study comes from a report in the *New Scientist* magazine in 1988, which first drew America's attention to the causes of coastal flooding and erosion. The data in Figure 79 refer to net movements of land in relation to sea level and show a clear pattern of change with a distinctly geographical distribution.

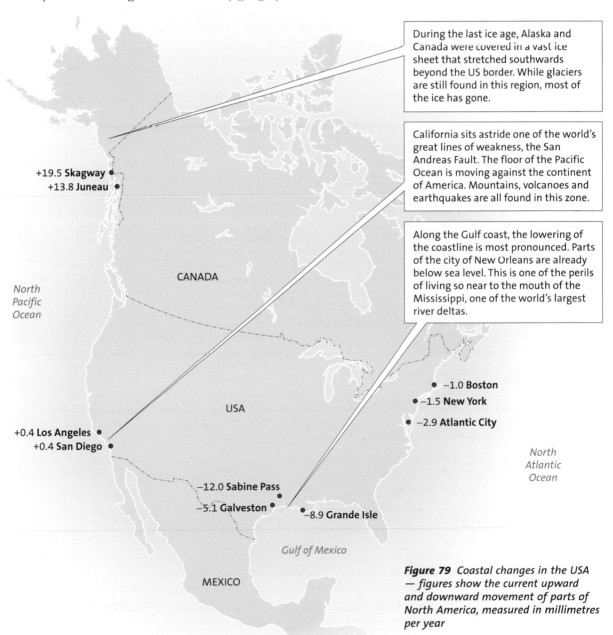

During the last ice age, Alaska and Canada were covered in a vast ice sheet that stretched southwards beyond the US border. While glaciers are still found in this region, most of the ice has gone.

California sits astride one of the world's great lines of weakness, the San Andreas Fault. The floor of the Pacific Ocean is moving against the continent of America. Mountains, volcanoes and earthquakes are all found in this zone.

Along the Gulf coast, the lowering of the coastline is most pronounced. Parts of the city of New Orleans are already below sea level. This is one of the perils of living so near to the mouth of the Mississippi, one of the world's largest river deltas.

+19.5 **Skagway**
+13.8 **Juneau**

*North Pacific Ocean*

CANADA

USA

+0.4 **Los Angeles**
+0.4 **San Diego**

−1.0 **Boston**
−1.5 **New York**
−2.9 **Atlantic City**

*North Atlantic Ocean*

−12.0 **Sabine Pass**
−5.1 **Galveston**
−8.9 **Grande Isle**

*Gulf of Mexico*

MEXICO

**Figure 79** *Coastal changes in the USA — figures show the current upward and downward movement of parts of North America, measured in millimetres per year*

## Question

Study the four groups of data shown in Figure 79.

(a) For each group, identify:
   (i)  its geographical location
   (ii) the general change in land level
   (iii) the causes of the change

(b) Choose two groups of data and explain the changes involved (some are more difficult than others).

## Guidance

For part **(a)** you may find it useful to give the answer in table form.

In part **(b)** the East coast is the easiest to explain, while the Gulf coast is the most difficult. Causes are linked to eustatic, isostatic and tectonic events.

# Landforms

The landforms resulting from sea level change are submergent or emergent.

### Submergent coastlines

This is when sea level change has been positive. A number of distinctive landforms and features result from this inundation of the land (Table 9).

| Landform or feature | Locations | Formation |
|---|---|---|
| Lowland estuary | • Suffolk<br>• Essex | • When the ice age ended, sea level rose<br>• Mudflats and tidal creeks were created<br>• Still prone to inundation during storms |
| Submerged forest | • Borth beach, Wales | • As the climate warmed, marsh and forest grew, but continued submergence drowned them<br>• Only tree stumps remain |
| Island chain | • Frisian Isles, Netherlands<br>• Croatian coastline along the Adriatic Sea | • Coastal hills submerged leaving a line of islands along the shore<br>• Mountainous coast even more spectacular |
| Ria | • South Devon<br>• Brittany<br>• Southwest Ireland | • Funnel-shaped, drowned, lower valleys of rivers along hilly coasts<br>• Now partly filled by deposition |
| Fjord or sea loch | • Western Scotland<br>• Norway<br>• South Island, New Zealand | • Long, deep, steep-sided inlets formed by submergence of previously glaciated valleys in mountainous terrain |

**Table 9** Features of coastal submergence

**Fjords** are found along mountainous coastlines where major glaciers once carved into the landscape. The deep, vertically sided troughs they created are found almost entirely in west-coast locations in middle to high latitudes, for example in Scotland, Norway, Canada, Chile and New Zealand. The milder westerly airstreams and higher temperatures of the lower altitudes encouraged the glaciers to melt at the seaward end of these valleys. The result was that the ice was much thinner and erosion less effective.

Today, with the ice gone and the valley flooded by the sea, a fjord has a relatively shallow sea entrance or threshold, which contrasts with the great depths further inland. Along the valley sides waterfalls fall from hanging valleys, while at the upper end of the fjord, material eroded by today's streams is deposited in the loch. This delta of new material is a welcome area of flat land and a valuable site for settlement.

The calm, deep waters of fjords make sheltered anchorages or sites for fish farms. Fjordland in South Island, New Zealand — which has some 14 fjords reaching up to 40 km inland — extends from Milford Sound in the north (Figure 80) to Preservation Inlet in the south.

*Figure 80 Milford Sound*

Jane Buekett

### Emergent coastlines

**Emergent coastlines** result from a negative change in sea level. We have already seen that land may recover after being depressed by ice sheets. In this situation, former coastal features may be uplifted and the sea retreats, showing land that was previously below sea level. Today's marine processes are in this way creating a new shoreline. Old cliff lines and other relict features are therefore found further inland and higher up. Raised beaches and fossil cliffs are two such landforms (shown in Figure 81).

- Raised beaches are seen along much of the coast of western Scotland and consist of shingle or shell deposits often on what was an old wave-cut platform. It is difficult to interpret the altitude of these features in relation to sea level changes as they seem to occur at various heights (8m, 15m and 30m). Nevertheless, they show that land has emerged from the sea and that this process has not been at a steady rate. Shells allow accurate carbon dating. Notable examples are found on Arran and in Gruinard Bay.
- Fossil cliffs are often found on the landward side of raised beaches. The spectacular King's Cave (Figure 82) is a series of caves cut in sandstone cliffs on the west coast of Arran. The pillars have notches where they were once cut by waves perhaps 6000 years ago. The raised beach in the foreground is made up of many large-calibre, rounded deposits, and is now covered in vegetation.

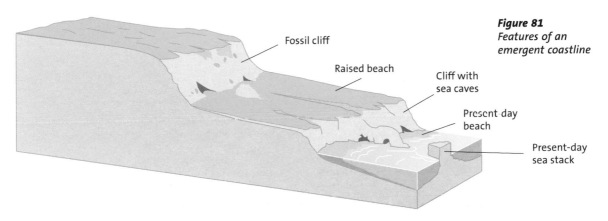

*Figure 81*
*Features of an emergent coastline*

Fossil cliff

Raised beach

Cliff with sea caves

Present-day beach

Present-day sea stack

**Figure 82**
*King's Cave, Arran*

*Martin Junius/www.m-j-s.net/photo*

## 36 Question

*Using case studies*

(a) Copy and complete the following table by writing the names of eight landforms in the correct places. Two ecosystems have been inserted already.

| Advancing shorelines | | Retreating shorelines | |
|---|---|---|---|
| Deposition | Emergence | Submergence | Erosion |
| Sand dunes | | Salt marsh | |
| | | | |
| | | | |

(b) For each landform in the table, identify the case study you would use to help explain its formation. You can use the same case study more than once.

(c) Look back at all the photos in Part 7 and identify as many landforms as you can.

## Guidance

These are brief tasks designed to help you classify landforms and understand the processes involved in their formation.

*Note: Although it is convenient to separate landforms and processes into discrete topics to help understanding, this is not the ideal approach. Not only does it create a rather stereotypical view, it ignores the fact that these processes rarely take place in isolation. Erosion, transportation and deposition often occur together, as seen in Case study 17. The coastline of western Scotland has both fjord features and raised beaches, indicating that both submergence and emergence have occurred in the same place but at different times.*

# Coastal sand dunes

Sand dunes form along most parts of the UK's coast, although there are fewer of them in the southeast. The main requirement for their continued formation is a fresh supply of sand. The ideal environment is one with a large tidal range and net deposition, which allows sand from river and marine sources to be re-worked by onshore winds. Dunes are dynamic and fragile environments within which an increasingly diverse ecosystem evolves. Two underlying concepts here are **succession** and **zonation**. Succession is about changes over time, while zonation is about how the ecosystem changes spatially away from the shoreline. Managing sand dunes is of increasing importance against a background of rising sea levels and coastal economic development.

## Succession

Succession refers to the changing structure of a plant community over time as it responds to environmental conditions such as shelter, soils and water. Starting from an initial abiotic environment, **pioneer plants** begin to invade and colonise the area. As they change the environment around them, these plants help the dunes to collect sand. They also provide shelter, which allows other vegetation to compete. Gradually, soil conditions and drainage change and at each stage the species involved succeed those already there. Each step in this process is called a **seral stage**. The eventual stage is known as the **climatic climax** because it is fully adapted to the climate and environment at that time. In sand dunes, this final stage is often heath, scrub or even woodland. As plants develop, they cover more of the bare ground and the variety of plants increases, leading to greater diversity.

However, this process can be interrupted. If the arresting factor responsible is natural, such as flood or storm damage, the stage is referred to as a **sub-climax**. If the cause is human, such as allowing grazing or trampling, then **plagio-climax** is the correct term to use. **Blowouts**, where sand has been removed and the dunes breached by wind erosion, may be either physical or human in origin. These may be the main cause of dune slacks where the water table is at or near ground level. In the sand dune ecosystem (**psammosere**), this drive to maturity is the **primary succession**. If progression is arrested and re-starts, this may lead to **secondary succession** (Figure 83).

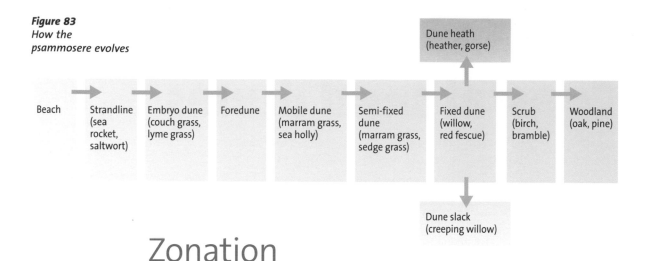

**Figure 83**
How the psammosere evolves

Beach → Strandline (sea rocket, saltwort) → Embryo dune (couch grass, lyme grass) → Foredune → Mobile dune (marram grass, sea holly) → Semi-fixed dune (marram grass, sedge grass) → Fixed dune (willow, red fescue) → Scrub (birch, bramble) → Woodland (oak, pine)

Dune heath (heather, gorse)

Dune slack (creeping willow)

# Zonation

Zonation refers to the way in which ecosystems develop to cover an area of ground. This is the spatial equivalent of succession, and in the case of dunes means a series of 'strips' of vegetation running parallel to the shore. Dunes appear as ridges and hollows, each occupied by dominant plants adapting to the changing environment. The basic pattern of dune zones can be seen in Figure 84.

Dunes begin when the movement of sand inshore is halted by an obstacle such as beach litter. Plants can also speed up this process, although they have to be tolerant of the high pH, salt spray and strong winds (Figure 85).

Of all the plants in the dune ecosystem, marram grass is perhaps the most remarkable, illustrating how successful species are adapted to survive in hostile environments (Figure 86). In mobile dunes, this plant copes with little shelter, limited water and temperatures reaching 60°C. There are few nutrients in the sand but the plant survives because of rhizomes and long stems, which can extract water from deep below the surface. The plant grows rapidly (1 m per year), avoiding damage if periodically buried by sand. Shiny, rolled leaves reduce evaporation. If the supply of fresh sand is interrupted, red fescue or creeping willow may replace marram as the dominant species.

**Figure 84**
Zonation in a transect across sand dunes

**Foredune:** an object or plant causes sand to build up on the lee side

**Mobile dune:** plants bind sand together

**Semi-fixed dune:** dunes could be 20 m high here

**Dune slack:** sand is removed by the wind until the damp sand near the water table cannot be transported

**Scrub, heath and woodland:** climax vegetation in the absence of management/interference

Marsh plants

Sandy beach

Sea

**Figure 85**
*Saltwort is an early invader to the foredune*

**Figure 86**
*Marram grass is well adapted to mobile dunes*

**Table 10**
*The structure of sand dune zones*

Bob Hordern

| Component | Dominant species | pH | Humus (%) | Colour | Commentary |
|---|---|---|---|---|---|
| Strandline | Sea rocket, saltwort | 8.5 | <1 | Yellow | Almost abiotic environment: dry, salty and few nutrients. Pioneer plants collect sand (Figure 85). |
| Embryo dune | Couch grass, lyme grass | 8.0 | <1 | Yellow | Succession starts. Sand accumulates and first perennial grasses colonise. Small dunes build up. |
| Mobile or foredune | Marram grass, sea holly | 7.5 | <1 | Yellow | First real humus results from marram decay. Shelter is an important factor and species diversity begins. |
| White, yellow or semi-fixed dune | Marram grass, sedge grass | 7.0 | 2.5 | Yellow | Mosses cover bare ground between marram grass colonies (Figure 86) and various non-grasses appear. |
| Fixed dune | Willow, red fescue | 6.5 | 5 | Yellow/grey | 'Fixed' implies no bare ground. Remains a fragile environment from blowouts, trampling or grazing. Now rich in flowering species. |
| Dune slack | Creeping willow | 6.5 | 10 | Yellow/grey | Controlled by the water table. Some distinct species present, e.g. orchids, natterjack toads. |
| Dune heath | Heather, gorse | 6.0 | 20 | Brown | Plagio-climax stage caused by grazing. Soils are dry, acidic and with limited nutrients. |
| Dune scrub | Birch, bramble | 6.0 | 20 | Grey | Natural succession leads to small trees and shrubs but the community becomes species poor. |
| Woodland | Oak, pine | 4.5 | >40 | Brown/black | Climax vegetation follows on from the scrub stage. The area is much at risk from human activities. |

**37** *Using case studies*

## Question

(a) What is the difference between succession and zonation?

(b) Give examples of how sand dune plants and their locations are affected by:
  (i) microclimate    (ii) soil conditions    (iii) other factors

## Guidance

(a) It is important to realise the difference — many students confuse these terms.

(b) Microclimate involves exposure and shelter as well as temperatures. Soil and water conditions are key factors in zonation. Other factors are related to people's use of dune areas.

## MORFA HARLECH: DIVERSITY IN THE DUNES

Set against the backdrop of Snowdonia, beneath Harlech Castle, Morfa Harlech is one of Britain's few actively growing sand dune systems. A large number of species of rare plants and birds have been recorded here. The old cliff shoreline where the castle is built is now some way inland, indicating just how much coastal deposition has taken place in the last 700 years (Figure 87).

The dunes at Harlech have most of the elements of a typical psammosere. These include high dunes, a major blowout, wet slacks and dune heath. The beach and frontal dunes are popular in the summer, so trampling and erosion take place, although control measures have been put in place (Figure 88). Golf and tree planting compete for space.

Sand dunes are flourishing at Morfa Harlech for two reasons:

- Deposition is rapid. Offshore sea bed deposits (glacial moraine) are brought ashore by waves and other material is eroded further south and brought by longshore drift. The large tidal range gives the sand time to dry out, so prevailing winds can move the loose sand up the beach (saltation).
- Isostatic effects (page 74) mean that the area is rising relative to sea level and this helps form a wide, flat beach that can be swept easily of its loose sand by strong winds. New dunes can form closer to the shore, ensuring succession.

**Figure 87**
*Harlech dunes from the castle*

Many students visit this area to investigate plant zonation. Most set up a number of transects, sampling vegetation cover and diversity using quadrats. Soil and

David Holmes

microclimatic data are recorded too. The results can be plotted on kite diagrams to give a pictorial view of vegetation patterns across the dunes (Figure 88).

Diversity means the range of different species of plant found. This also shows a pattern across the dunes and tells us more about how zonation works. The number of species found (species richness) is a crude way to measure diversity as it makes one

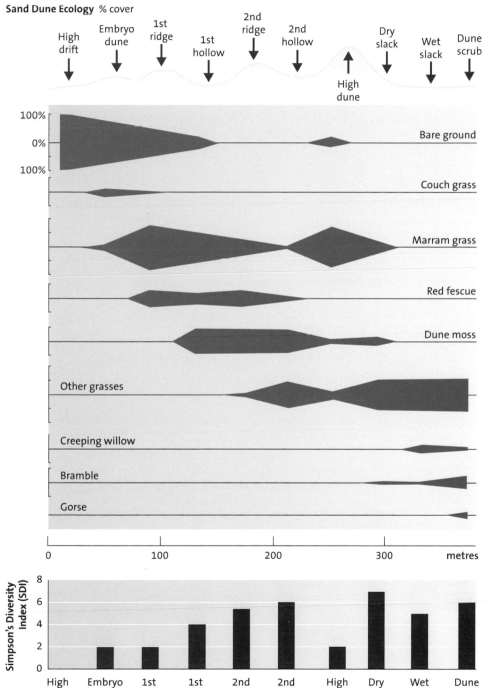

**Sand Dune Ecology** % cover

*Figure 88*
*Kite diagram of percentage cover and graph of diversity across Morfa Harlech dunes*

daisy as important as a thousand buttercups. For this reason, we use Simpson's Diversity Index (D), which takes into account both species abundance and diversity:

$$D = \frac{\Sigma n(n-1)}{N(N-1)}$$

where $N$ is the total number of organisms of all species and $n$ is the total number of organisms of a particular species.

Further details of the sand dunes at Morfa Harlech and Aberffraw on Anglesey are available on the GeoResources website (**www.georesources.co.uk/csdintro.htm**).

## 38 Question

**(a)  Why is Morfa Harlech a good place to study succession and zonation?**

**(b)  Using Figure 88, describe and explain the pattern of vegetation occurrence and diversity across the dunes.**

### Guidance

Comment on trends and then illustrate your points with a few species/locations.
For example:
*The marram grass needs fresh sand to flourish, so it is confined to areas where the wind is strong. It is therefore more common on ridges (80% on the first ridge) than in hollows. Salt spray damages the leaves, so it is not found on the beach. Competition from other species like grasses and creeping willow prevents its growth inland.*
To use Simpson's Diversity Index, go to **www.countrysideinfo.co.uk/simpsons.htm**.

# Managing threats

Threats to sand dunes can be natural — as with some blowouts — but most are human. Figure 89 shows the range of threats involved.

*Figure 89
Threats to sand dunes*

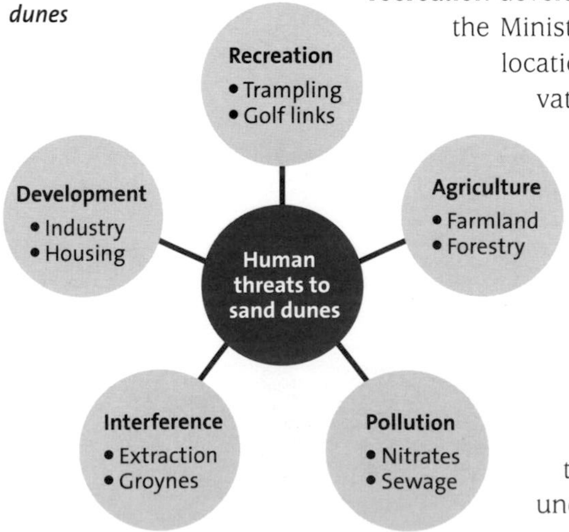

Conflicts are common because there are many interested parties. Land ownership, recreation developments, grazing rights, forestry, sewage works and even the Ministry of Defence seem to be common opponents in these locations. Not surprisingly, English Nature and other conservation groups also have a viewpoint.

**Recreation**
• Trampling
• Golf links

**Development**
• Industry
• Housing

**Agriculture**
• Farmland
• Forestry

**Human threats to sand dunes**

**Interference**
• Extraction
• Groynes

**Pollution**
• Nitrates
• Sewage

Managing sand dunes is a complex issue and, as with other coastal situations, decisions about the degree of interference or conservation are difficult. Traditional responses, such as afforestation, stabilise shifting sand, but this can prevent dune migration. Introducing new species can also be counterproductive, lowering water tables and reducing species diversity. Preventing grazing by rabbits or sheep may help. The key is to get a balance where there is still some instability — but not too much. An integrated and consultative approach is undoubtedly the best strategy.

David Holmes

**Figure 90**
Dune management
at Harlech

## AINSDALE DUNES

*Case study* **21**

The Ainsdale dunes on Merseyside are an outstanding area of wildlife interest with a range of conservation and scientific designations. The fixed dunes and dune slacks provide habitats that protect the rare sand lizard and natterjack toad. English Nature manages the Ainsdale Nature Reserve within the wider Sefton Coastal Management Scheme. Threats to the area's 7 km² of dunes come from natural and human sources:

- Visitors make heavy use of the area; there are 5 million people living within one hour's drive. This leads to erosion and habitat loss, mainly from litter, fires and trampling.
- Non-native species, like pine, poplar and grasses, are invading the dunes from the landward end of the site (Figure 91).

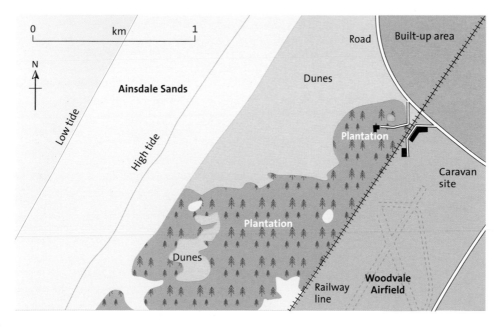

**Figure 91**
A sketch map of the
Ainsdale dunes

There are a number of management questions to be resolved:
- Should the sand dune complex be restored and preserved, removing pines?
- Should the pinewoods be retained to protect the red squirrel population?
- Should managers let nature to take its course?
- Should a variety of ecosystems be managed by careful intervention or just the dunes?

Strategies might include:
- visitor management — make some areas closed or allow permit-holders only into the dunes nature reserve
- species management — scrape pools for toads to breed, fence off lizard areas
- scrub cutting — cut small areas only as tractors are too damaging
- mowing — this controls the height of creeping willow or invaders
- turf stripping — this restarts succession in old fixed dunes
- grazing — allow sheep and cattle to graze the creeping willow in fixed dunes
- dune restoration — make some attempt to stabilise the dunes if necessary

For further details of the Ainsdale dunes, visit the Sefton Coast website (**www.sefton-coast.org.uk**). The Liverpool Hope University College website (**www.sandsoftime.hope.ac.uk**) also provides useful information and is an excellent source for revision.

## Question

**Management of the Ainsdale sand dunes is generally successful. Which strategies do you think make this possible?**

### Guidance

To investigate the Ainsdale dunes in a different way, go to the Multimap website (**www.multimap.com**) and type in Ainsdale in the search dialogue box. Change the map scale to 1:25 000. Click the camera icon to change the view to an aerial photo. Move the curser over the dunes area in the bottom left to see an overlaid map. Use the pan arrows and zoom feature to help you identify detailed features of the dunes, beach and plantation.

# People and coasts

Coastlines present us with a range of opportunities that are fast becoming a global problem as people continue to demand to live, work and holiday there. In many countries, the coast is overcrowded, overdeveloped and over-exploited.

- In 2000, 40% of the world's population was living within 100km of a coastline, which is only 20% of the world's land area.
- 50% of Americans live within 6km of the sea or the Great Lakes; 75% within 100km.
- China, the USA and Brazil have the fastest-growing coastal urban populations.
- In Europe, the Mediterranean coast is under the greatest pressure, with as many tourists as residents in summer.
- Mumbai in India is responsible for putting a million tonnes of sewage into the Indian Ocean every day.
- 50% of Indian Ocean coral reefs have died since 2000.

What is the attraction of the coast? Initially, settlement along coasts allowed exploitation of marine resources such as fish and oil. Commerce and international trade led to the growth of ports and major coastal cities. In modern times, beaches, attractive scenery and diverse habitats have encouraged a growing recreation and tourism industry.

The interaction of people and the environment gives rise to many coastal issues. Figure 92 summarises interactions at the coast and forms a good basis for identifying case studies that reveal coastal issues.

People–environment issues at the coast include flooding, changing deposition, coastal erosion, economic development, population pressure, resource depletion, ecosystem loss, pollution and global warming. All are pressing issues and many are interdependent. Two examples are outlined below:

- In 1991, cyclones hit the coast of Bangladesh — a country threatened more than most by global warming — causing large-scale flooding and costing 140 000 lives. The mangrove forests, endangered by development of the Sundarban islands in the delta area, prevented coastal erosion and further loss of life.
- In the Caribbean, development of the coasts and oil resources is increasing pollution levels, which are damaging

*Figure 92*
*Environmental interaction at the coast*

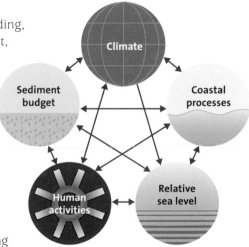

coral reefs. These reefs are an important aspect of the tourist industry, which is the mainstay of the economies of many of the region's islands.

This section considers issues relating to short-term processes, such as rapid coastal erosion and floods, as well as the longer-term effects of sea level change. As the economic development of coasts continues, the role of marine ecosystems will become more important. Tough decisions about coastal defences will need to be taken. Inevitably, this leads to conflicts of interest as decision-makers struggle to find cost-effective, yet environmentally acceptable, solutions.

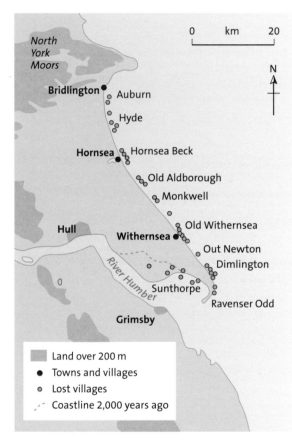

# Rapid coastal erosion

Coastal erosion along the Yorkshire coast between Bridlington and Spurn Head results from a combination of storm waves, longshore drift and a relatively weak coastal geology. There is evidence that 29 settlements have been lost since Roman times (Figure 93).

Perhaps more worryingly, surveys suggest the rate of erosion may be accelerating. Whether this results from natural changes, global warming or local human interference remains to be seen.

Evidence of this erosion process can investigated at individual points along the coast using the Hull University website (**www.herb.hull.ac.uk/erosion/index.htm**), which overlays aerial photos with an OS map of 1852.

Natural cliff erosion at the Holderness coast happens via the following cycle:
- the soft boulder clay cliffs become saturated with rain water

**Figure 93**
*The lost villages of East Yorkshire*

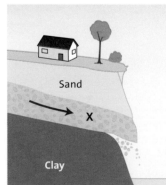

**Figure 94**
*Human causes of cliff instability*

- the steep cliff face 'fails' and a landslide or slump takes place
- the cliff and debris create a relatively stable angle of debris
- storm waves remove the debris in longshore drift, the cliff steepens and the cycle begins again

Building on the cliffs adds to the problem as it not only increases loading on the clay but also interferes with drainage and increases overland flow (Figure 94).

## 40 Question

### Using case studies

**Study Figure 94.**
**(a) Identify and explain the sub-aerial processes taking place at points X and Y.**
**(b) Suggest how cliff-top development may have influenced these processes.**

### Guidance

These questions focus on how cliffs are affected by terrestrial (land-based) processes as well as marine processes. Refer to Part 2 for more technical information.

## MEMORIES OF MAPPLETON

*Case study* **22**

The coastal village of Mappleton has changed considerably over the last century (Figures 113 and 114 on pages 110 and 111). Mappleton's problem, like many of its neighbouring settlements, is its boulder clay cliffs and narrow beach.

Figure 95 shows the changes at Mappleton between 1910 and 1988. While the settlement has grown and new buildings have been built, the biggest change is the loss of land. You can also see how the beach has narrowed. Buildings along Cliff Road were already threatened by 1988, and the coastal road near Gray's Farm would soon be at risk.

*Figure 95*
*Changes to the village of Mappleton, 1910–88*

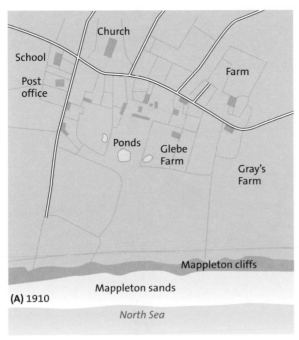

**(A)** 1910

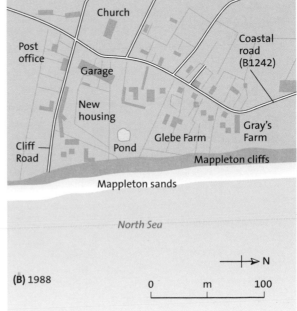

**(B)** 1988

## Question

Using Figure 95 and Figure 114 on page 111, track the changes in the beach, cliffs and buildings in 1910, 1988 and 2001.

## Guidance

Consider the relative positions of the various features. Try to use the scale to assess the threats to buildings and people.

# Coastal flooding

The effects of storms on coasts can be dramatic, causing rapid erosion or flooding. However, the risk of overtopping or breaching of defences in low-lying areas causes most alarm. The hurricane coast of the USA is to some extent protected by its barrier islands, whereas the cyclone-prone delta area of Bangladesh has fewer natural defences. In the UK, the storms of January 1953 flooded much of the east coast of England when a storm surge caused by a deep depression coincided with high spring tides. Around 300 people died, together with 2000 more in the Netherlands. Many of the world's coastlines are low-lying and numerous islands, particularly in the tropics, have little or no high ground.

*Case study* **23** COASTAL FLOODS IN TOWYN

Towyn is a holiday resort near Rhyl on the North Wales coast, facing the Irish Sea. On 26 February 1990 the town suffered a major flood event (Figure 96). Waves overtopped the sea wall at 10.30 a.m. and began to flood the low-lying land behind. By lunchtime, the wall — which carried the coastal railway line — was breached and sea water inundated over 10 km² of Towyn and the land around it. An evacuation of 5000 people took place from houses, bungalows and retirement homes. Holiday camps were ruined, and services and power supplies badly disrupted.

**Figure 96**
**Impacts of coastal flooding, Towyn, 1990**

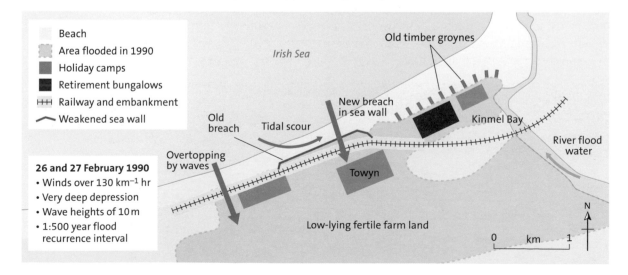

There were several reasons why this event became a hazard:

■ A deep depression (951 mb) drove 130 km$^{-1}$ hr winds onshore
■ This intensity of storm could not have been foreseen (its recurrence interval was calculated as perhaps once in 500 years).
■ The storm coincided with very high tides.
■ The groynes were old and had allowed the beach to degrade.
■ The sea wall had to carry the railway as well as protect the shoreline.
■ Housing and bungalows had been built on low-lying land behind the wall.
■ Many of the residents were elderly people living on their own.

## 42 Using case studies

### Question

Using *Case study 23*:
(a) identify the physical causes of the flooding
(b) explain how this hazard might have been prevented or had its impact reduced and suggest what steps local agencies might take to protect the coastline in the future

### Guidance

(a) Physical causes relate to the effects of the storm and the sea conditions.
(b) Human influences may have included previous management decisions or land-use planning.

# Tropical coastal ecosystems and small islands

Many tropical islands are made of, or protected by, coral, or have coastal mangrove ecosystems. These are threatened by natural short-term events, such as cyclonic storms, and by the long-term implications of a global rise in sea level. Their low relief is a further complication (Figure 97).

Marine scientists suggest that 60% of the world's coral reefs have been damaged by human activity, many beyond recovery. Most reports blame pollution, coastal development and farming, all of which can damage the fragile ecosystems that make up coral reefs. In addition, many changes in land use, such as forest clearance

(A) 'Celebrity Love Island', Mamanuca Fiji

(B) Reefs, the Whitsunday Islands, Australia

Fiona Flood

**Figure 97**
*Islands and coral reefs at risk from storms and rises in sea level*

*Figure 98*
*A coastal ecosystem scorecard*

**Scorecard**

| Food/fibre production | Water quality | Biodiversity | Recreation | Shoreline protection |
|---|---|---|---|---|
|  |  |  | ? |  |

**Key**

**Condition** assesses the current output and quality of the ecosystem good or service compared with output and quality of 20–30 years ago

| | Excellent | Good | Fair | Poor | Bad | Not assessed |
|---|---|---|---|---|---|---|
| Condition | | | | | | |

**Changing capacity** assesses the underlying biological ability of the ecosystem to continue to provide the good or service

| | Increasing | Mixed | Decreasing | Unknown |
|---|---|---|---|---|
| Changing capacity | ╱ | ↑↓ | ╲ | ? |

| Goods | Services |
|---|---|
| • Fish and shellfish | • Protection from storms |
| • Fishmeal (animal feed) | • Wildlife habitat |
| • Seaweed (food, industry) | • Biodiversity |
| • Salt | • Human settlement |
| • Genetic resources | • Employment |
| | • Recreation/tourism |

*Table 11*
*The value of coastal ecosystems*

and intensification of agriculture, cause increased erosion. This washes soil into rivers and out to sea, silting up coastal reefs and cutting off the sunlight they need. Clearance of mangroves for tourism and fishing is also increasing.

Coastal ecosystems are valuable resources in terms of the goods they supply and the services they perform (Table 11). The World Resources Institute in Washington DC carries out a health check on the condition of these ecosystems. The state of coastal ecosystems in 2002 is shown in Figure 98.

# Threats to coral reefs

- Many reefs have been poisoned by pollution. Sewage and runoff contaminated with fertilisers damage reefs by increasing the nutrient levels in sea water, allowing seaweed to outpace coral growth.
- Coral reefs have been destroyed by mining (for building materials) or burning (for lime production). Coral items are sold as souvenirs.
- Tourists are blamed as many reefs have been damaged by divers, walkers, anchors and overfishing.
- On a wider scale, higher water temperatures and increasing carbon dioxide bleach the corals and make the water more acidic.

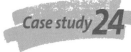
**Case study 24** **ST LUCIA UNDER STRESS**

St Lucia, a holiday destination in the Caribbean, is what a topical island paradise should look like. But beneath the surface there are threats to its future. These threats come from both physical and human causes, but human actions are potentially more damaging.

In September 1994, tropical storm Debbie struck St Lucia. Continuous rain (400 mm in 10 hours) led to soil erosion, massive landslides and widespread flooding. What made matters worse was that the island had carried out much deforestation for agricultural use and a new west coast road was being excavated. Headlines in the newspapers told of the damage caused but little was said about the silent damage to the coral reefs that happened at the same time.

### Question

(a) How valuable are tropical coastal ecosystems?

(b) Summarise why people are to blame for the damage to coral reefs.

(c) Why might the 'silent' damage to corals be worse than the storm damage that made the headlines?

### Guidance

Refer to Figure 98 and Table 11 when answering these questions. Look at the 'scores' for coastal ecosystems; what goods and services do they provide?

## Physical and human factors

The most obvious question about people and coasts is whether physical or human factors have most effects. Studies by the EU Commission confirm that, in Europe, human interference in coastal systems creates the most impacts, the scale of which are shown in Figure 99.

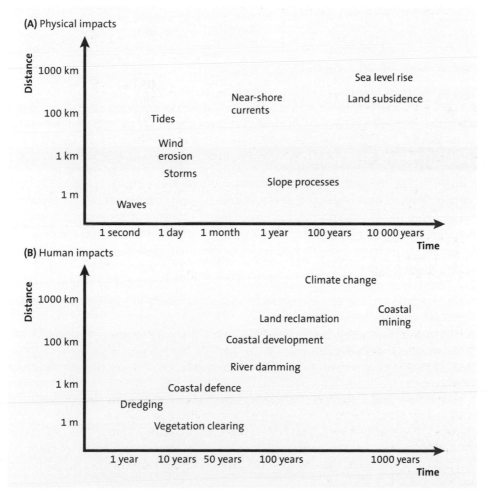

*Figure 99*
*Impacts of physical processes and human activities on coastal environments*

**Question**

**Read the following statements. Can you add three more statements for each category?**

■ Physical factors: waves have an immediate impact whereas storms last for days.

■ Human factors: climate change is global and its effects are long lasting.

**Guidance**

To answer this question, refer to Figure 99. The *x*-axes estimate impacts over time while the *y*-axes give an idea of the scale of their spatial impacts.

# Coastal conflicts

Conflicts arise along coasts often because economic development creates environmental consequences. Different groups or individuals have their own views on proposals to resolve coastal conflicts. They might be financially linked to the issue, politically or legally responsible, or directly affected by it. They might be experts in it or represent a particular lobby. Decisions can be arrived at relatively easily, or they may lead to a public enquiry or drag on inconclusively over many years. Conflict/compatibility matrices, collective bargaining agreements (CBAs) or critical path analysis can be involved in the decision-making process.

Two of the most common conflict scenarios relate to either:

■ concerns about a specific issue or decision (to be taken or one taken previously)

■ general disagreements about a whole range of issues (usually about land use)

*Case study 25*  **PROBLEMS IN PORTLAND BAY**

Portland is near the southern tip of Australia in the state of Victoria. In the 1960s, new harbour facilities were developed, which included the construction of two breakwaters. Initially, this allowed the port to function more effectively but the main breakwater deflected waves and problems soon became apparent, because of longshore drift:

**1** Deposition appeared along the main breakwater and it became necessary to dredge the entrance to the harbour.

**2** A sand trap was built and material from there was transported by road to nourish the beach beyond the lee breakwater. But beaches in Portland Bay were soon starved of sand, which accelerated erosion.

**3** Residents at Henty Bay complained until a rubble sea wall was built to protect the houses and an access road.

**4** Subsequent changes have led to erosion (terminal scour) of the next section of coast down drift.

**5** Sand is now accumulating further along the shoreline, inland of Minerva Reef.

Conflicts about who was to blame and what to do about the problem soon became big local issues:

■ Residents felt that the port authority and politicians who wanted the harbour development were to blame, and demanded they build further defence works.

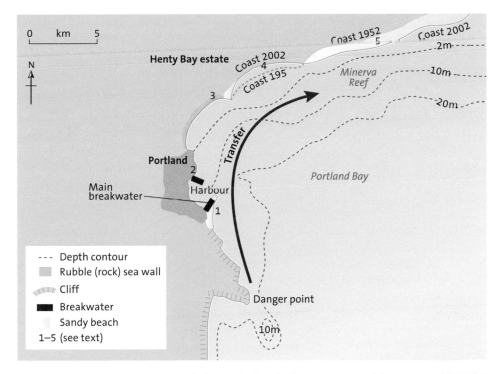

**Figure 100**
*Portland Bay*

- Conservationists were worried about the level of environmental damage and felt the problem had simply been shifted to Minerva Reef.
- Local politicians argued that the breakwaters were necessary to protect investment in the deepwater port and the aluminium smelter, which now provides 800 jobs and earns A$80 million each year.
- The port authority needed the port to handle imports of alumina and oil, and exports of cereals, woodchips and livestock.
- Local businesses were originally linked to port trade but today there is more tourism, sports and conservation throughout the whole Portland Bay area.

## 45 Question

**Using case studies**

(a) Outline the costs and benefits of building the harbour breakwaters at Portland Bay.

(b) Explain the conflicts that exist and why each group holds its beliefs.

### Guidance

Use the text and Figure 100 to answer both questions. Resolving conflicts is important in coastal management, as you will see in Part 10.

## CONFLICTS IN NORTHUMBERLAND

*Case study* **26**

At the southern end of the Northumberland coastline, between the towns of Blyth and Newbiggin, is the Wansbeck estuary. To the north is Sandy Bay where, because of coastal defences at Newbiggin, the cliffs are eroding rapidly. To the south, deposition dominates, caused by southward longshore drift.

Use of this land along this coast has created conflicts of interest (Figure 101). A legacy of heavy industry from mining and manufacturing has left its mark. Newer businesses are chasing tourist spending and conservationists wish to see the coastline return to its pre-industrial beauty.

*Figure 101*
*Land use conflicts along part of the Northumberland coastline*

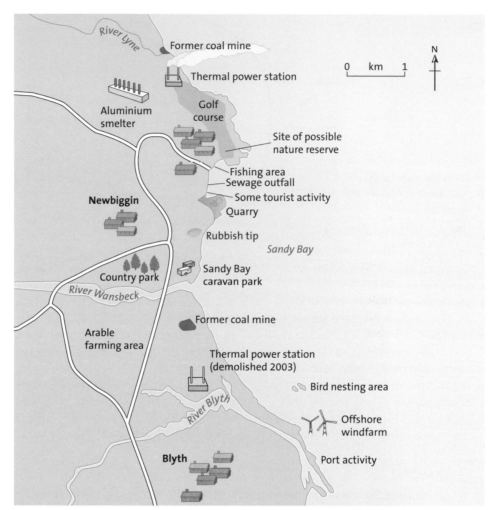

**Question**

(a) **Suggest why this coastline is affected by widespread pollution.**
(b) **Identify and explain two possible land use conflicts.**

**Guidance**

To answer these questions, look at Figure 101. The map shows the conflicts involved. Sandy Bay is a classic coastal study area in many post-16 textbooks.

# Coastal management

Managing coastlines is about reducing the effects of erosion and flooding, preventing unwanted deposition, protecting people and their property, and managing the way we use areas alongside the coast.

Defending coasts involves a variety of groups, including local councils, DEFRA and the Environment Agency. English Nature and the National Trust may also have strong interests in particular locations. These groups provide a forum for discussion and cooperation and play an important role in the development of **shoreline management plans (SMPs)**.

## Shoreline management plans

The objectives of SMPs are to improve our understanding of coastal processes and to set up the long-term planning of coastal defences and land use. An SMP 'assesses the risks to people and the developed, historic and natural environment, while devising a sustainable strategy for flood and coastal defence'. First-generation SMPs are currently being upgraded to meet future needs.

Since beginning the first round of these management plans, it has become obvious that two other factors are of increasing concern. These are the effects of global warming (rising sea levels and increasing storminess) and the need to deliver sustainable solutions (financially, socially and environmentally). SMPs are now in place for all 11 sediment cells in England and Wales (Figure 102) and details of these can be found on the internet The Regional Coastal Defence Group's website (**www.coastalgroups.co.uk/coastal/shoreline.html**)

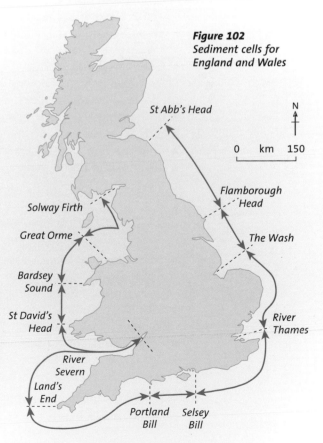

**Figure 102
Sediment cells for England and Wales**

St Abb's Head

Flamborough Head

The Wash

Solway Firth

Great Orme

Bardsey Sound

St David's Head

River Thames

River Severn

Land's End

Portland Bill

Selsey Bill

0   km   150

is a good starting point with useful links. Scotland has nine areas and the Scottish National Heritage website (**www.snh.org.uk**) is a good starting point for researching these.

In practice, these large cells are sub-divided into smaller units to allow careful examination — for example, Flamborough Head to Spurn Head is Unit 2a. This sub-cell is broken down further to allow appropriate local decisions to be made.

For mini summaries of SMPs and defence strategies, the Eurosian Portal website at **www.eurosion.org** provides snapshots of most of Europe's coastlines and the decisions taken.

## Related schemes to research

- Integrated Coastal Zone Management Plans are similar to SMPs but often include civic and business aspects. Details of the East Riding of Yorkshire plans can be accessed on the East Yorkshire Coastal Observatory website at **www.hull.ac.uk/coastalobs/index**. This excellent site, run by Hull University, has an interactive map, details of coastal erosion and management strategies, 3600 panoramas and video flyovers at sites along the coast. There are also links to other relevant websites.
- The Standing Conference of Problems Associated with the Coastline (SCOPAC) site at **www.scopac.org.uk** provides details of coastal defence schemes along the central south coast. Click on the 'Maps' link and then take the 'Coastal Defence and Protection' option. When the map appears, use the 'Zoom in' tool to focus on particular coastal defences.

## Choosing an SMP

Once an erosion or flooding issue has been identified, four broad options are available to decision-makers:

- **do nothing** — literally take no action other than monitor the situation
- **retreat** — pull back, setting up a new line further inland
- **hold the line** — maintain or sustain the present line of defence
- **advance the line** — build forward (this is a rare decision)

The following assessments are applied to any planned change:

- **Cost–benefit analysis** divides benefits by cost to give a ratio. Capital and maintenance costs are weighed against benefits such as property protection, employment and more difficult measures like people's safety.
- **Environmental impact assessment** looks at the effects on the environment of a particular scheme. Impacts on sediment movement, damage to habitat, risk of pollution and loss of heritage are some examples.
- **Feasibility studies** consider the technical aspects of a plan, involving an understanding of the factors and processes involved, and the engineering design.
- **Risk assessment** considers aspects such as the recurrence interval of storms or flooding events and how long defences are designed to last. It then weighs these against the value of what is at risk — particularly people and property.

The decision about which option fits in with the management objectives can be explored using a matrix that scores each option against general and specific criteria.

| Option | | Do nothing | Retreat | Hold the line | Advance the line |
|---|---|---|---|---|---|
| General objectives for whole coastline | Technically feasible | ✔ | ✘ | ✔ | ✔ |
| | Economically viable | ╱ | ✘ | ✔ | ✘ |
| | Environmentally acceptable (built) | ✘ | ✔ | ✔ | ✔ |
| | Environmentally acceptable (natural) | ✘ | ✘ | ✔ | ✘ |
| | Compatible with process | ✔ | ╱ | ✘ | ✘ |
| | Compatible with adjacent units | ✔ | ✔ | ✔ | ✔ |
| | Sustainable | ✘ | ╱ | ╱ | ✘ |
| Specific objectives | Protect town and activities | ✘ | ● | ✔ | ✔ |
| | Maintain sediment movement | ✔ | ╱ | ● | ✘ |
| | Maintain EU bathing quality | ✘ | ✔ | ✔ | ✔ |
| | Maintain RNLI and fishing access | ✘ | ✔ | ✔ | ✔ |

**Key**
✔ Meets objectives
✘ Conflicts with objectives
● Meets some, but not all, objectives
╱ Insufficient data

Decision: hold the line

*Figure 103
A decision matrix
for Hornsea, East
Yorkshire*

The example shown in Figure 103 is for Hornsea, a resort on the Holderness coast, where the decision taken was to 'hold the line'. If you count the 'ticks' in Figure 103, you will see that this policy scores most, and is the only one to meet economic and environmental criteria.

Remember, this is about *policy*, not techniques or methods.

It is often not easy to obtain up-to-date figures for cost–benefit analyses, nor is it straightforward to put a value on people's sense of security or on business confidence. However, some simple comparisons can be made. For example, a 14-million coastal defence scheme at Shoreham near Brighton seems good value if it protects £135 million worth of property (1300 homes and 90 businesses) for 100 years, whereas Happisburgh's plans on the Norfolk coast for cliff stabilisation and breakwaters costing several million Euros will exceed the falling value of the 18 houses and road currently at risk. For more information about Happisburgh's plans, visit the Happisburgh Village website at **www.happisburgh.org**.

# Coastal management techniques

Having decided on a suitable SMP, the next step is to choose the most appropriate defence methods from the wide variety of techniques available. These vary in cost, durability and effectiveness. Much depends on making an appropriate choice in relation to the physical processes occurring, the use of the land and the risks involved. Such methods are usually divided into hard and soft techniques, although their use may be part of a wider sustainable or integrated management plan.

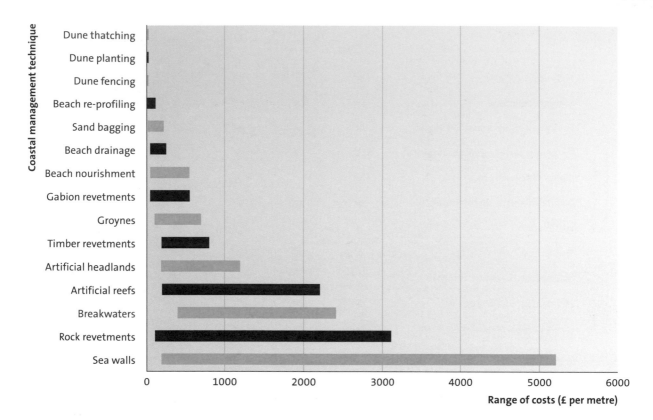

**Figure 104** *The range of costs of some coastal management techniques, 2000*

Some defences are more expensive than others, but it is difficult to establish average costs. Scottish Natural Heritage has attempted this and produced a budget sheet of costs per metre for 15 common techniques, from which mean values can be calculated (Figure 104).

# Hard engineering

Table 12 describes a number of hard engineering management techniques.

The hard defences at Withernsea on the Yorkshire Coast, shown in Figure 105, cost £6.3 million. They consist of a combination of different techniques.

- The groynes along the beach are the first line of defence. They trap much of the sand moving south (away from the camera) in the longshore drift. This holds the beach in place, helping to slow the speed of the waves as they break along the coast. This is a relatively low-cost method.
- At the top of the beach, a newly built sea wall made of concrete, is the last defence. This protects the promenade and the hotel buildings behind, from heavy seas in the winter months. The shape is carefully designed to deflect, rather than resist the waves. This is an expensive solution to coastal erosion.
- In front of the wall, engineers have placed rip-rap of rock armour. These heavy boulders, imported from Scandinavia, not only break up the waves but also prevent the sea wall from being undermined by wave scour.

| Technique | Operation | Limitations |
|---|---|---|
| Breakwaters | Offshore structures of concrete or rocks that deflect waves before they reach the shore; the term can also refer to protective walls near harbours | Tend to deflect waves along new paths, which may cause scouring of foundations or erosion elsewhere |
| Gabions | Metal cages filled with rocks, which can be stacked to build walls (often 1 m cubes); may also be arranged as a mattress so that waves can percolate | Metal cage may fail relatively quickly, spilling contents; if stacked, they may move in strong waves |
| Geotextiles | Permeable fabrics that accumulate materials yet allow water to pass through; modern types (geotubes) are filled with dredged sand | Relatively new and there is limited evidence of results on a large scale |
| Groyne fields | These wooden or rock barriers, sited perpendicular to the shore, trap and retain drifting sand; repairable; the key is to get a balance of sand stored and sand still moving | Likely to interfere with the sediment budget along a coastline, causing deficit or even starvation downdrift |
| Revetments | Sloping ramps (concrete, rock or open wooden baffles) to absorb wave energy; modern versions have large blocks placed in a finer-mix foundation | Despite sloping design, revetments have similar problems to sea walls; they may damage foreshore ecosystems |
| Rip-rap (rock armour) | Similar to sea walls, but permeable and able to adapt to change; larger rocks are not easily moved; can look relatively natural depending on location | Rocks can roll in severe weather or be under-scoured by backwash; not always visually attractive and usually expensive |
| Sea walls | These bulkheads separate land from water, supporting the land as well as holding back the sea; modern designs can absorb and deflect as well as resist wave attack; seen as appropriate, if not mandatory, in promenade and harbour situations | Wave return walls may accelerate beach erosion due to downward motion of water and scour at toe of wall; often the most expensive option both in terms of capital and maintenance costs |

It is this combination of methods that provides Withernsea with an effective defence. This 'belt and braces' approach will increase the life of the engineering structures and reduce maintenance costs. In a wider economic sense, the protection of the hotels and the improvements to the beach will help retain tourism, which is vital for a small holiday resort.

*Table 12*
*Hard engineering coastal management techniques*

Bob Hordern

*Figure 105*
*A combination of different types of hard defence at Withernsea*

## Case study 27 · BLACKPOOL: STILL A HARD CASE

Conditions along the Lancashire coast vary from the shallow, strongly tidal Morecambe Bay in the north to the more exposed drift-aligned features of the southern Sefton coast. In general, the land is low-lying and flooding rather than erosion is the major issue facing planners. The central section, the Fylde, contains a number of popular holiday resorts, so it is not surprising that the usual policy is to 'hold the line'.

Blackpool — with its promenade, piers, tower and pleasure beach — has used massive sea walls and concrete structures to protect its considerable tourist investment. Capital costs and repairs are expensive. Winter storms overtop the promenade and wave scour undermines foundations. Despite this and the apparent fall in tourist income recently, the resort has received £62 million of funding through DEFRA to develop the central promenade. The new sea front will protect 1500 business and residential properties. The promenade area will be increased by 5 hectares and access to the beach will be improved. It is hoped that this 'Masterplan', which will be completed by 2009, will also regenerate the economy of the town.

**(A)** Old sea wall

**(B)** New sea wall

Bob Hordern

**Figure 106**
*Sea walls, Blackpool*

Figure 106 shows how hard engineering design techniques have moved from wave resistance to deflection and absorption. Only to the south of the resort are softer management techniques used, taking advantage of the sand dune ecosystems there.

# Soft engineering

Other types of coastal defence are known as soft because they attempt to work with natural processes rather than control them. One benefit of such schemes is that they are generally less expensive to implement. A negative aspect may be that they can involve loss of property or land. Table 13 outlines some of the soft engineering techniques available.

| Technique | Operation | Limitations |
|-----------|-----------|-------------|
| Beach drainage | Drainage improvement for swash-aligned beaches using coarser sand; reduces surface backwash and gives better beach colour (for beach users) | A new concept not tested fully; will not work on drift-aligned beaches |
| Beach nourishment | Artificial replacement of sand from land or water (pumped onshore as slurry) | Exogenous (imported) sediments may have environmental and visual drawbacks |
| Beach reprofiling | Economical way to to resist erosion by changing the shape of the beach | 'Scraping' only works in low-energy environments; storms can deconstruct the shape |
| Cliff drainage | Reduces saturation by piping water out of cliffs and so prevents landslips | Only applicable to some rock types |
| Cliff profiling | Slope of cliff face lessened (regraded) to improve stability (critical angles involved) | Requires detailed research into the geology of the cliffs |
| Cliff toe protection | Blocks placed at cliff foot beneath potential failure sites | Largely a temporary response, as this method does not halt erosion |
| Creating stable bays | Increasing length of bays in an attempt to spread wave energy along the coast and focus wave erosion on headlands; protects bays | Not yet tried in Europe; long-term approach |
| Dune regeneration | Wind velocity reduced by fences or planting above beach level to encourage sand deposition; generally effective if managed properly | Effective only in aeolian environments and where public access can be controlled |
| Marsh creation | Planting mudflats with pioneer species like spartina to increase stability in tidal flow; seen as an affordable and perhaps sustainable solution | Technique may be jeopardised by an accelerated rise in sea level if marsh plants cannot adapt quickly enough |
| Mudflat recharge | Similar to beach nourishment, but uses cohesive sediments (e.g. in Dutch Polders) | Similar limitations to marsh creation |
| Rock-pinning | Bolting rock layers to prevent slipping along cliff faces | Only applies to some cliffs and does not prevent wave erosion |
| Sand bypassing | Removal and re-injection of sediments into a sediment cell; used to bypass important harbour breakwater features, which keep navigation channels clear | Despite common use (Case study 25), the costs are becoming increasingly prohibitive |
| Vegetation planting | Use of planting to stabilise cliffs and dunes (e.g. kudzu and marram grass) | These fragile environments need careful management |

# Sustainable coastal defences

*Table 13*
*Soft engineering coastal management techniques*

Sustainable coastal defences are relatively new to the UK. They try to accommodate, copy or work alongside natural systems and processes, with ecosystems often playing a key role. These techniques are at an early stage in development and are often only small in scale, remedial and bottom up (locally driven). As an approach to defence, they are environmentally friendly and may offer a longer-term solution for many places along the coast. Many people feel that hard engineering strategies no longer provide a reliable or cost-effective solution to coastal defence, so the case for alternative, sustainable techniques gets stronger.

Studies of Blackpool (*Case study 27*) and Mappleton (*Case study 30*) have highlighted the economic costs and environmental impacts of defending coasts, while Towyn (*Case study 23*) is a reminder that management has to cope with flooding hazard as well as rapid erosion. Reasons for a change in policy are as follows:
- Global sea level continues to accelerate.
- Rapid erosion is resulting from increased storminess.

- Our coastlines are 'fixed' by development and cannot change ('coastal squeeze').
- Much of the existing coastal engineering is still relatively low technology.
- Coastal ecosystems of all types are under pressure globally (salt marshes, sand dunes, corals and mangroves).
- In human terms, coastlines are too popular, overused and overdeveloped.

One of the most discussed types of sustainable defence is **managed retreat**. This allows the sea to flood parts of the intertidal zone of coastal lowlands rather than people spending vast sums of money defending areas that will ultimately become marine. Such schemes have important ecological credentials as they can create mudflats and valuable salt marsh habitat. This is especially important because 50% of such areas has been lost in the UK in the last 30 years.

Land trapped between the rising sea and aging sea walls is in a 'coastal squeeze', especially in southeast England. **Coastal realignment** is controversial as it involves 'retreating the line', which means no longer defending some parts of the Essex, Kent and north Norfolk coastlines. Salt marsh ecosystems are able to trap material and, through the process of succession, create new land and habitats. In normal circumstances, they can adapt to small increases in sea level.

Figure 107 shows how the situation around the North Sea coast has changed over time with the realisation that coastal defences have to respond to natural and human developments:

- In Stage 1, farming along the coast is based on livestock and cattle graze the pasture land and extensive salt marshes.
- By Stage 2 (largely begun after the Second World War), demand for food sees the pasture replaced by arable production and perhaps draining the marshes. At the same time, dykes/walls are built to protect the farmland from coastal flooding. The area of saltmarsh and habitat has reduced in size.

**Figure 107**
*The changing face of North Sea coasts*

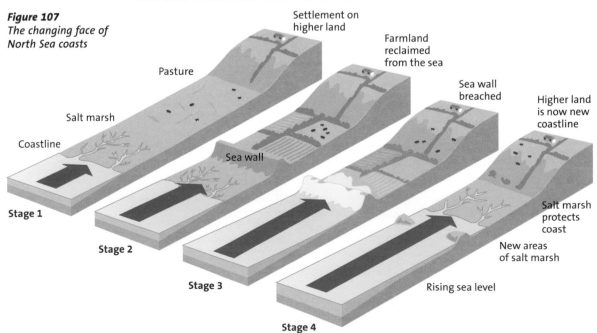

Settlement on higher land

Pasture

Salt marsh

Coastline

Farmland reclaimed from the sea

Sea wall

Sea wall breached

Higher land is now new coastline

Stage 1

Stage 2

Stage 3

Stage 4

Salt marsh protects coast

New areas of salt marsh

Rising sea level

*Contemporary Case Studies*

- Stage 3 shows the situation in recent years.
- Stage 4 shows the way forward: managed retreat

## ESSEX GOING SOFT

A eustatic rise in sea level and an isostatic subsidence of the land is having a double effect on tidal shorelines in Essex. Salt marsh and mudflats are eroding at the rate of 2 m per year, which is faster than at Holderness. As a habitat, the area is rich in wading birds. The mudflats provide their food supply and the salt marshes their nest sites. The river estuaries form natural nurseries for fish.

Abbotts Hall Farm at Great Wigborough in the Blackwater estuary was chosen as the site of an 80 hectare managed retreat scheme, which saw the breaching of a section of the sea wall. The scheme, the largest in Europe in 2002, was supported by the Environment Agency, Colchester Borough Council, WWF, English Nature, Essex Wildlife Trust, farmers and local oyster fishermen. The spring tide was allowed to flood the land on 4 November.

In order to obtain planning permission, a full environmental assessment was carried out and the estuary was monitored before and after the flooding. Many fauna were carefully moved or protected, and bird feeding and nesting times were avoided. The arable farm itself will generate income to support the educational work of the Essex Wildlife Trust, and plans can be seen in Figure 108.

More information can be found in the September 2003 issue of *Geography Review* or on the WWF marine update at **www.wwf.org.uk/filelibrary/pdf/mu52.pdf**.

**Figure 108**
**Map of Abbotts Hall Farm**

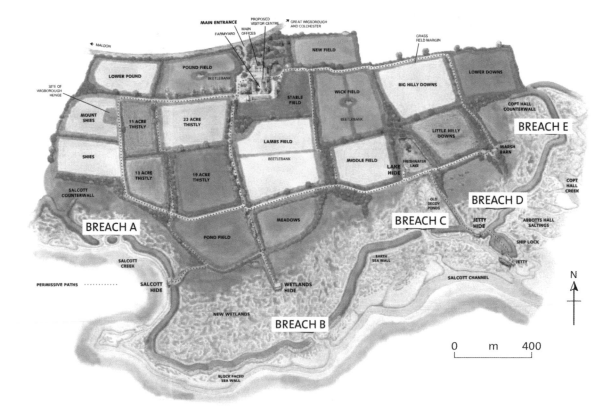

## Question

**(a)** Study Figure 107.
   **(i)** Explain why the situation in Stage 3 is considered to be unsustainable.
   **(ii)** Suggest reasons why the managed retreat strategy in Stage 4 is:
   ▪ sustainable
   ▪ especially suited to this type of coast

**(b)** Choose two of the environmental groups mentioned in this case study and suggest why each one supported the scheme.

## Guidance

**(a)** When answering this question, consider first what happened in stages 1 and 2. Ask yourself why the management methods must change.

**(b)** Note that the support for Abbotts Hall Farm may be economic or environmental.

# Shoreline management plans in action

When planning to defend the coastline, there are situations when hard defences are necessary, particularly in well populated resorts. There are other areas where less valuable land uses or important habitats suggest a softer approach. Sustainable solutions also need to be considered whenever possible.

In practice, coastal managers select a number of different strategies along a coast to meet particular needs. While one type of option or technique may dominate, there has to be flexibility over the proposals.

*Case study* **29**

## AN SMP FOR HOLDERNESS

Erosion is the main issue along this cliff coast and it is both rapid and continuous. Landslips, wave quarrying and beach erosion are the main processes at work along much of the coast. Population densities are generally low and agriculture is the most common land use. Therefore, the general policy is to 'do nothing'.

However, at six specific locations, including four small holiday resorts, 'hold the line' is the chosen policy. Easington, with its gas terminal, is also being protected, as is the north shore of the Humber estuary where flooding is an added risk.

A 'do nothing' policy has been suggested for the Spurn peninsula, but there are conflicts over this decision. Figure 109 summarises the current situation.

More information can be found on the East Yorkshire Coastal Observatory website at **www.hull.ac.uk/coastalobs/ index** and **www.hull.ac.uk/coastalobs/ media/pdf/iczm.pdf**.

A summary of the policies and techniques is shown in Table 14. The options to advance or retreat the line have not been implemented on this coast. It would be valuable to compare defence techniques in place at Barmston, Hornsea and the Spurn peninsula.

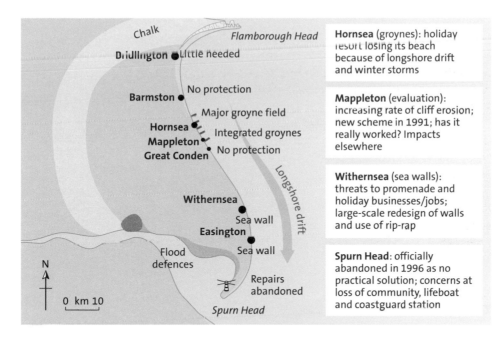

*Figure 109*
*Managing the Holderness coast*

## Barmston: 'do nothing'

The three villages of Barmston, Atwick and Skipsea are all inland of the retreating boulder clay cliffs. Historical documents show the speed of this erosion and record the loss of the earlier settlement of Attenwick. Today, farmland and caravan parks are the main land use between the coastal road and the sea. Recent losses at Barmston (population 200) have been the wooden steps that allowed access to the beach. A café and caravan site may be next. Temporary defences have included some tipping of rocks to protect the cliffs from the waves, but the 'do nothing' policy means that this will soon end.

*Table 14 Summary of coastal defence decisions, 2005*

| Sub-unit | Name/location | Policy | Technique/method | Comments |
|---|---|---|---|---|
| 1 | Flamborough | Do nothing | | Chalk headland; resistant to erosion |
| 2 | Bridlington | Hold the line | Promenade, walls, groynes | Main resort; little erosion of beach |
| 3 | Fraisthorpe | Do nothing | | Farmland |
| 4 | Barmston/Atwick | Do nothing | Private gabions at Skipsea | Small villages inland; monitor |
| 5 | Hornsea | Hold the line | Promenade, walls, groynes | Holiday resort (£5.2 million spent on defences) |
| 6 | Rolston | Do nothing | | Farmland |
| 7 | Mappleton | Hold the line | Rip-rap, rock groynes | Small village; main coastal road |
| 8 | Aldbrough/Tunstall | Do nothing | | Farmland |
| 9 | Withernsea | Hold the line | Promenade, walls, groynes | Holiday resort (£6.3 million spent on defences) |
| 10 | Holmpton | Do nothing | | Farmland |
| 11 | Easington Terminal | Hold the line | Rip-rap, perhaps until 2020 | Main gas terminal (£4.5 million spent on defences) |
| 12 | Easington/Kilnsea | Do nothing but retreat periodically | | Small village inland; monitor |
| 13 | Spurn peninsula | Do nothing but local retreat or intervention possible | | At-risk nature reserve, SSSI and coast guard station |
| 14 | Sunk Bight | Hold the line | Flood bank | At risk from flooding; farmland |
| 15 | Sunk Island | Hold the line | Flood bank | At risk from flooding; farmland |

**Figure 110**
*A groyne trapping sediment even in rough sea, Hornsea*

Bob Hordern

At nearby Skipsea, one local caravan park owner has invested in some hard engineering, building a series of **gabion cages** reinforced with concrete to try to prevent winter wave erosion. Evidence suggests this will fail as the coast on either side is eroding rapidly, rather like a headland.

### Hornsea: 'hold the line'

This town of 8000 people is dependent on tourism and some local fishing. The 2.9 km shoreline is lined with small hotels and a promenade. Groynes have been the traditional defence but sand losses have been occurring and the coast to the south of the town has suffered increasing erosion from terminal scour (Figure 110).

To 'hold the line', new defences were built in 1994, which included the replacement or repair of 19 groynes, underpinning sea walls, building a flood wall and placing **rip-rap** in front of the low south promenade. The design of these works was tailored deliberately to avoid damage to the beach front or deter holidaymakers and commercial

**Figure 111**
*A summary of management issues at Spurn*

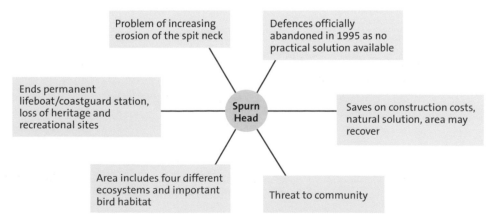

Problem of increasing erosion of the spit neck

Defences officially abandoned in 1995 as no practical solution available

Ends permanent lifeboat/coastguard station, loss of heritage and recreational sites

**Spurn Head**

Saves on construction costs, natural solution, area may recover

Area includes four different ecosystems and important bird habitat

Threat to community

operators. Pedestrian access to the beach is provided using steps over the low sea wall. Steel gates allow fishermen's vehicles to reach the sea but prevent low-level flooding by spring tides. Erosion to the south has reduced and new housing has even been built close to the coast. The total cost has been £5.2 million.

## The Spurn peninsula: 'do nothing/local retreat or intervention'

Spurn is a 5 km recurved spit (*Case study 18*) that is migrating slowly southwestwards into the Humber estuary. This may be part of a natural long-term cycle, but the problem is how to manage events now.

| Option | | Do nothing | Retreat |
|---|---|---|---|
| Human objectives | Coastal defence | ✗ | ● |
| | Fishing | ✗ | ✔ |
| | Tourism | ● | ● |
| | Offshore use | ✔ | ✔ |
| | Navigation | ✗ | ✔ |
| Physical objectives | Biology | ✔ | ● |
| | Natural processes | ✔ | ● |
| | Landscape | ● | ● |
| | Water quality | ✗ | ✔ |

Key
✔ Meets objectives
✗ Conflicts with objectives
● Meets some, but not all, objectives

Decision: do nothing

*Figure 112*
*A decision matrix for Spurn*

The decision to 'do nothing' is not quite what it seems; there are reservations about this in the long term, and 'local retreat or intervention' is anticipated. The downdrift impacts of the rip-rap at Easington Terminal are a new factor to consider. The long-term decision is made more difficult by the presence of a nature reserve, a Site of Special Scientific Interest (SSSI) and a coastguard station. To the north and west, Sunk Island and Sunk Bight are both currently protected by flood banks, but neighbouring Kilnsea is likely to 'retreat the line periodically'.

'Managed retreat' would probably make Spurn a sand and shingle island rather than a spit, the breach taking place in the narrow neck area. Historical and geomorphological evidence suggesting the spit will simply migrate relates to times when there was relatively little interference in processes along the Holderness coast — a situation that is not the case today.

## 48 Question

(a) What factors have led to the 'hold the line' policy at various locations along this coast?
(b) Why is 'do nothing' the most popular option in terms of kilometres of coastline?
(c) Why are sea walls (or rip-rap) and groynes the most used techniques on this coast?
(d) Study Figures 111 and 112. Why was the decision at Spurn Head 'do nothing'?

### Guidance

Most of these questions relate to land use and whether locations are farmland or holiday resorts (see Table 14). In part **(d)**, Figure 112 suggests that 'retreat' should be the proper option. How does Figure 111 help explain the planners' decision?

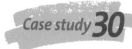 

Although cost–benefit and environmental impact analyses allow engineers and consultants to try to gauge the effects of their plans before implementing them, it is only afterwards that the real, long-term impacts become clear. This is especially true with schemes that attempt to deal with longshore drift problems because the likelihood of affecting locations downdrift is considerable. Mappleton on the Holderness coast is a good example. This case study will evaluate its shoreline management plan.

■ Historical records show how the village and its predecessors had progressively been eroded by strong wave attack and landslips triggered by rainwater mobilising the boulder clay cliffs. In 1786, the church was 3.5 km from the sea, 200 years later, the sea was on the doorstep, with the beach inaccessible and houses on Cliff Road collapsing.

■ Residents and local pressure groups campaigned to be given some coastal defences and eventually this happened, although the decision to protect the village was taken primarily to secure the B1242 coast road (Figure 95 on page 89). The management decision was to 'hold the line' and to try to tackle the processes at the foot of, and on the face of, the cliffs.

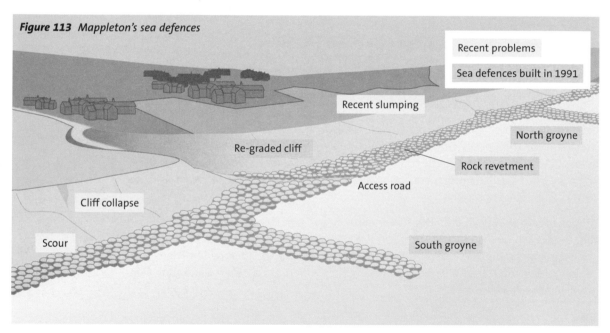

**Figure 113** Mappleton's sea defences

Recent problems

Sea defences built in 1991

Recent slumping

North groyne

Re-graded cliff

Rock revetment

Access road

Cliff collapse

Scour

South groyne

The scheme adopted had three components, all funded with a £2.1 million EU grant (Figure 113):

■ Two rock groynes were constructed, designed to catch and build up sand as it moved southwards in suspension along the coast.

■ A rock revetment (rip-rap) was placed along the foot of the cliffs to counteract wave erosion in storm and spring tide conditions.

■ **Re-grading** of the boulder clay cliffs was carried out to reduce the risk of landslips and at the same time to allow a new road access to the beach.

The main photo in Figure 114 shows an aerial view of the Mappleton defence scheme in 2001, looking southwards following the direction of longshore drift, with the beach apparently stabilised and the coast road saved. However, the smaller photos indicate places where perhaps all is not well:

- A: there are early signs of landslips/slumping in the re-graded cliff profiles.
- B: the cliff by the car park has been undercut by terminal scour.
- C: cliff retreat has reached alarming rates to the south at Great Cowden.

**Figure 114**
*The Mappleton defence scheme and some of its impacts*

**49** **Question**

*Using case studies*

(a) **What reasons would have led planners to choose the 'hold the line' option?**
(b) **Suggest why engineers chose the three components at Mappleton.**
(c) **Explain why some experts are now critical of the defence scheme.**

**Guidance**

This question is exploring three main topics:
(i) the reasons for defending Mappleton
(ii) the techniques chosen
(iii) the success or otherwise of the scheme after 10 years

# Managing coasts in LEDCs

In less economically developed countries (LEDCs), decisions about shoreline strategies and techniques are generally more restricted, in terms of both finance and technology. Perhaps because of this, a 'do nothing' approach is often the only realistic response to coastal issues. While hard defences exist near ports and cities, many of these are relics from colonial developments.

The impact of the Akosombo Dam on the West African coast shows how difficult coastal management can be for countries with a range of problems to tackle, such as poverty, food production, HIV/AIDS and internal conflicts. It is also easy to see why economic factors outweigh environmental concerns.

However, the coastal ecosystems of many LEDCs (mangroves and corals in particular) could provide a sustainable way forward. This change in approach is being seen in Bangladesh — with its 'green river' proposals, which stretch into the delta region — and in countries like Guyana, where new coastal strategies are being developed as much by necessity as by choice. Countries where tourism is of key importance, such as those in the Caribbean and Indian Ocean, need sustainable strategies if they are to protect and retain both their resources and their customers.

## Case study 31 CHANGING COASTAL STRATEGIES FOR GUYANA

**Figure 115**
*Some physical and human aspects of Guyana's coastal areas*

Guyana is the only English-speaking country in South America. It is still largely covered by tropical rainforest, although much of the lowland area has been cleared to create plantations. There is obvious eco-tourism potential, but political instability, inter-ethnic tension and economic mismanagement have left it among the world's poorest countries, with an infrastructure that is barely able to support its population.

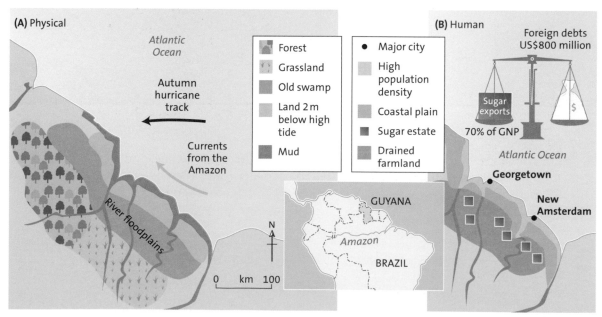

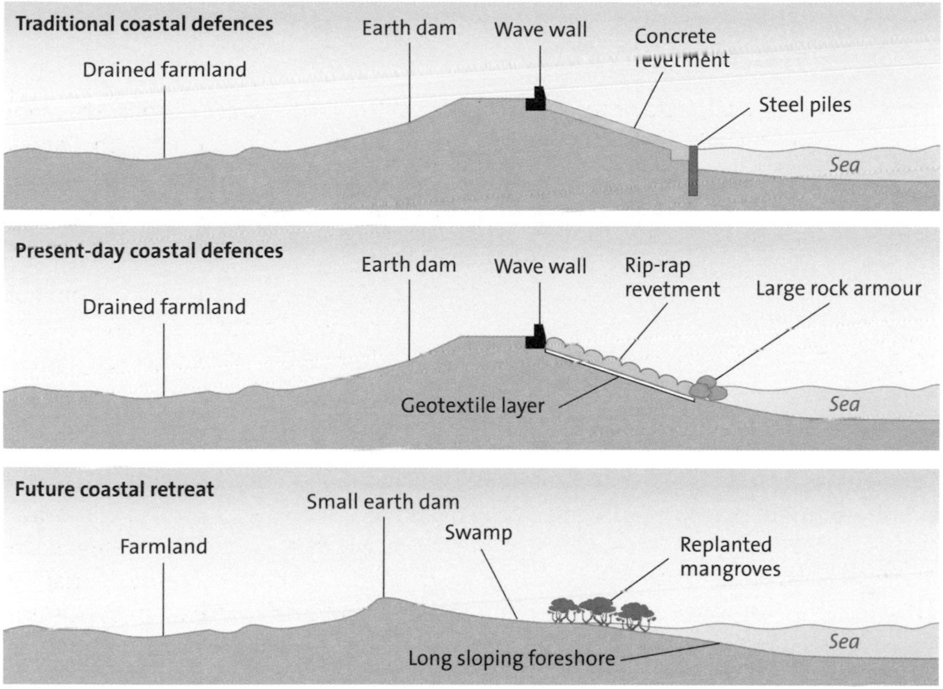

**Figure 116**
*Changing approaches to coastal defence in Guyana*

Parts of the coastal area are below sea level. In the past, Dutch settlers built an elaborate system of sea defences, using sea walls or dykes to keep out the sea, together with a network of canals, pumps and sluices. As Figure 115 shows:

- various physical factors make it difficult to protect and manage this coast
- a predicted sea level rise of 30cm by 2030 suggests a difficult future
- economic and social pressures are typical of many LEDCs

Guyana's coastal defences are in a poor state, with 50% beyond repair. By 2001, there was an immediate need for US$260 million of investment but Guyana responded to this by moving towards more sustainable coastal management techniques.

**50**

*Using case studies*

### Question

(a) Summarise the physical threats to Guyana's coast.
(b) Outline the arguments for and against Guyana spending one third of its income on coastal defence.
(c) Study Figure 116.
　(i) List the advantages of the traditional coastal defences.
　(ii) Suggest why new features of the present-day solution are an improvement to traditional defences.
　(iii) Examine the advantages and disadvantages of Guyana's future coastal plans.

### Guidance

When answering parts (a) and (b), use Figure 115 to explain the economic and environmental pressures on Guyana. Part (c) focuses on past, present and future defences. Examine each of the diagrams carefully and remember that Guyana is an LEDC.

# Examination advice

There are a number of ways to use case studies in an examination, depending on the wording of the question and the marks allocated to it. For example:

■ using **named examples** of what you are writing about

*The Holderness coast shows the effects of rapid coastal erosion…*

■ referring to **key facts and figures** that support your answer

*In terms of cost–benefit, a £14-million coastal defence scheme at Shoreham near Brighton seems good value if it protects £135 million worth of property (1300 homes and 90 businesses) for the next 100 years.*

■ using one case study to show **detailed knowledge and understanding** of a topic

*Mappleton on the Yorkshire coast shows how the use of hard engineering in one location can have effects further along the coast. The groynes built there have inter-rupted the southerly flow of sediment in the longshore drift and created a beach that will help to protect the cliffs from erosion. However, immediately beyond the second groyne there is clear evidence that…*

*A sketch map of Shrewsbury*

■ referring to named examples to show **wider knowledge and understanding**

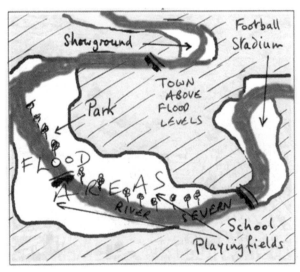

*The impacts of flooding on people's lives can vary greatly. In the delta region of Bangladesh, the sheer scale of events — in terms of time, area covered and lives lost — contrasts with those in Boscastle. Two months of flooding, two-thirds of the country inundated with water and 900 dead is a world away from Boscastle. It is easy to see flooding just as a hazard with negative and damaging effects on people and property. However, it should not be forgotten that in countries like India and China the annual floods have positive impacts: re-filling aquifers, allowing irrigation, flushing out pollution and spreading fertile silt across the floodplains.*

■ drawing a **diagram or map** to summarise or explain ideas

*Flood risk has an impact on land use, especially in urban areas. In Shrewsbury, planners have avoided building on the floodplain next to the River Severn. Instead, the land has been used as shown in the map.*

**Planning an essay** means using your case studies well. Let's look at one essay question in detail: 'For one named river catchment that experiences floods, examine the strategies used to manage flooding.'

- You should already be aware, from looking at previous papers, of the length of essay expected, the marks allocated, the time available and some likely essay titles.

- Work out what the essay is about: underline or circle the topic, key words and geographical terms. Using our example:

'For *one* named river *catchment* that experiences floods, *examine* the *strategies* used to *manage flooding.*'

'Strategies' could mean traditional structural/hard engineering or softer/more sustainable approaches. 'Examine' means you must do more than describe, so you could explain the strategies and say why they were chosen or evaluate them.

- Decide which case studies to refer to. Only one is allowed in our example, so which one? Top answers need either detail or to include a wide range of ideas.

- Plan how to structure your essay: introduction, main part (draw up a quick list or spider diagram) and conclusion.

- The introduction should be brief, but must mention catchment (not just river) and explain the term 'strategy'. Choose one river: the candidate below chose the Ouse in Yorkshire (*Case study 15*) although you may, of course, choose another. Build the essay around a range of strategies and refer to examples within the catchment. The conclusion should sum up what you want to say at the end: it is best to be brief.

- Here is one candidate's response. This is a top-grade answer — can you see why?

*The catchment of the River Ouse and its tributaries include high moorland in the Dales, flat floodplain in the Vale of York and tidal reaches in the Humber estuary. Such a large and varied river requires a range of strategies to manage flooding. Strategies to deal with floods can involve hard or soft engineering.*

*One way to examine these strategies is to categorise them into flood abatement, flood protection and flood adjustment. Abatement involves wider catchment strategies. In the Swale headwaters, there has been a lot of re-afforestation to intercept rainfall, and old drainage channels have been blocked up to reduce runoff. The National Park gives grants to plant woodland, such as at Nidderdale, where there are also storage reservoirs. Flood protection is often structural, including channelisation and building up levées. Both of these are used downstream of York, where the river is tidal. There is a flood barrier (earth embankment) between Selby and Cawood, while the river channel (deepened) and banks (reveted) have been modified within cities like York. The northern part of the city is protected by a washland area, the Ings, which is allowed to flood to save other areas. There is even a turnover floodgate in the city to prevent water from the Ouse backing up into the River Foss. Adjustment includes well-organised flood warning schemes and emergency plans. Some historic buildings have been flood-proofed and there are schemes to insure property. Land use along the banks is now controlled to prevent inappropriate development.*

*In the future, the River Ouse will still flood but the impacts are likely to be less damaging because, throughout the catchment, strategies have been used to allow people to live with, rather than try to control, flooding.*

Table 15 gives 20 essay questions that you can plan or write as part of your revision. A useful starter activity would be to fill in the last column for yourself.

**Table 15** *Essay titles*

| Essay title | Case studies | Figure/table references |
|---|---|---|
| Using named examples, explain how physical factors can influence either storm hydrographs or river regimes | 1, 2, 5 | |
| Use contrasting examples to compare the impacts of storms on river hydrographs | 1, 6 | |
| Referring to examples, explain how river landforms change downstream | 4 | |
| Using named examples, examine the factors that influence either river cross profiles or long profiles | 3, 4 | |
| Referring to examples, examine how and why river channel variables change with distance downstream | 5 | |
| Using examples, explain how human activities along rivers can lead to conflicts | 9, 10, 11 | |
| Using named examples, examine the costs and benefits of hard river engineering | 6, 11 | |
| Describe some examples of sustainable river management and explain their strengths | 12, 13, 14 | |
| For a named coastline, examine the effects of structure and lithology on coastal features | 16 | |
| Using examples, explain how transportation and deposition create coastal features | 17, 18 | |
| Referring to examples, examine how storm events can lead to either rapid coastal erosion or coastal flooding | 22, 23 | |
| Using examples, explain how changes in one named coastal ecosystem can result from both physical and human activities | 20, 21, 24 | |
| Explain how sea level changes can affect landforms along named coastlines | 19 | |
| Referring to a named coastline, examine how human activity can lead to conflict | 25, 26 | |
| Using examples, examine ways in which people can threaten and manage sand dunes | 20, 21 | |
| Referring to named examples, examine the view that hard coastal engineering creates more problems than it solves | 25, 27 | |
| Evaluate the success of coastal management along one coastline | 25, 30 | |
| Using examples, show how views about coastal management are divided | 27, 28 | |
| With reference to more than one country, describe factors that influence government coastal defence policy | 29, 31 | |
| Referring to one coastline, explain how human activities can lead to conflict | 26 | |

Maps and diagrams can also be useful for revision. Figure 117 shows an example drawn by a candidate, summarising the physical geography of the Holderness coast. This map could be compared with Figure 109 on page 107, which is a summary of the coastal management policies. Adapting a case study in this way in response to the actual question asked is a valuable skill.

*Contemporary Case Studies*

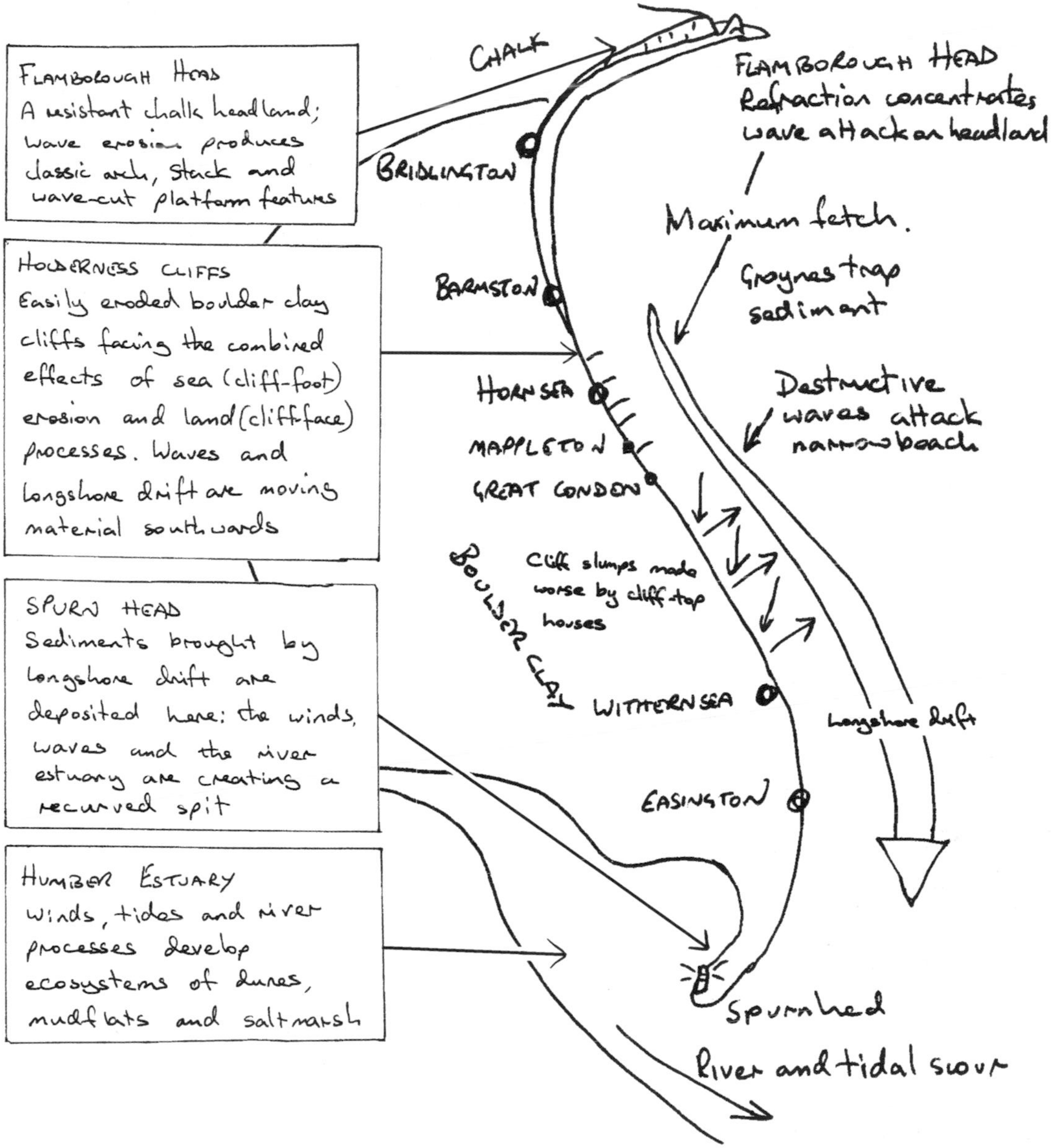

FLAMBOROUGH HEAD
A resistant chalk headland;
wave erosion produces
classic arch, stack and
wave-cut platform features

HOLDERNESS CLIFFS
Easily eroded boulder clay
cliffs facing the combined
effects of sea (cliff-foot)
erosion and land (cliff face)
processes. Waves and
longshore drift are moving
material southwards

SPURN HEAD
Sediments brought by
longshore drift are
deposited here: the winds,
waves and the river
estuary are creating a
recurved spit

HUMBER ESTUARY
winds, tides and river
processes develop
ecosystems of dunes,
mudflats and saltmarsh

CHALK
BRIDLINGTON
BARMSTON
HORNSEA
MAPPLETON
GREAT CONDEN
BOULDER CLAY
WITHERNSEA
EASINGTON

FLAMBOROUGH HEAD
Refraction concentrates
wave attack on headland

Maximum fetch.

Groynes trap
sediment

Destructive
waves attack
narrow beach

Cliff slumps made
worse by cliff top
houses

Longshore drift

Spurn head
River and tidal scour

*Figure 117*
*Sketch map of the
Holderness coast
showing physical
aspects of the coast*

# Index

*Contemporary Case Studies*